MODERN NUMERICAL METHODS
FOR ORDINARY DIFFERENTIAL EQUATIONS

MODERN NUMERICAL METHODS FOR ORDINARY DIFFERENTIAL EQUATIONS

EDITED BY

G. HALL AND J. M. WATT

CLARENDON PRESS · OXFORD

Oxford University Press, Ely House, London W.1

OXFORD LONDON GLASGOW NEW YORK
TORONTO MELBOURNE WELLINGTON CAPE TOWN
IBADAN NAIROBI DAR ES SALAAM LUSAKA ADDIS ABABA
KUALA LUMPUR SINGAPORE JAKARTA HONG KONG TOKYO
DELHI BOMBAY CALCUTTA MADRAS KARACHI

ISBN 0 19 853348 9

Printed in Great Britain

by J.W. Arrowsmith Ltd., Bristol

PREFACE

The aim of this book is to give practitioners of numerical methods in industry, research establishments and tertiary education a theoretical and practical account of the state of the art in modern numerical methods for the solution of ordinary differential equations. The book contains a full account of the best available algorithms together with a discussion of recent advances in theory, for both initial- and boundary- value problems.

Part I deals with the basic theory of methods for initial-value problems and contains a detailed description of the implementation of algorithms for the non-stiff case. In Part II stiff initial-value problems are considered. After a discussion of the special difficulties associated with stiffness there follows a description of methods in current use. Stiff differential equations arising in two important areas of application are considered in the final two chapters. Boundary value problems are taken up in Part III. After an introductory chapter on finite difference methods there is a full account of recent implementations of the shooting method. The application of the method to linear eigenvalue problems and generalized boundary-value problems, including handling of singularities, is treated in some detail. The last chapter describes expansion methods and the special circumstances in which they can be advantageously applied. The final section, Part IV surveys available methods in the recently developed areas of delay-differential and Volterra integro-differential equations.

This book is based on the material presented at a joint Summer School given in July 1975, organized by the Department of Computational and Statistical Science, University of Liverpool and the Department of Mathematics, University of Manchester. A full list of contributors follows this preface. We acknowledge with thanks the careful and skilful typing of the text, which was done by Miss B. Lee, Mrs. D. Manley and Mrs. J. O'Connor at Liverpool and Mrs. E. Bishop, Mrs. R. Horton and Mrs. K. Littler at Manchester.

G. Hall

J.M. Watt

CONTENTS

CONTENTS

CONTENTS

CONTRIBUTORS

Prof. J.C. Butcher (Chapters 5, 10)
Department of Mathematics, University of Auckland
Private Bag, Auckland, New Zealand.

Dr. J.D. Lambert (Chapter 2)
Department of Mathematics, The University,
Dundee DD1 4HN, Scotland.

Dr. A. Prothero (Chapters 9, 11)
Shell Research Ltd., Thornton Research Centre,
P.O. Box No. 1, Chester CH1 35H England.

Dr. H.H. Robertson (Chapter 13)
I.C.I., Corporate Laboratory, The Heath,
Runcorn, Cheshire, England.

Dr. C.T.H. Baker (Chapters 20, 21)
Dr. I. Gladwell (Chapters 12, 16, 17)
Dr. G. Hall (Chapters 6, 8)
Prof. J.E. Walsh (Chapters 16, 18)
Dr. J. Williams (Chapter 1)
Department of Mathematics, The University,
Manchester M13 9PL, England.

Prof. L.M. Delves (Chapter 19)
Dr. M.A. Hennell (Chapter 15)
Dr. R. Wait (Chapters 7, 14)
Mr. J.M. Watt (Chapters 3, 4, 15)
Department of Computational and Statistical Science,
The University, Liverpool L69 3BX, England.

PART 1

INITIAL VALUE PROBLEMS

1

INTRODUCTION TO DISCRETE VARIABLE METHODS

1. INTRODUCTION

Our objective is to discuss the theory and practice of numerical methods for solving the initial value problem

$$y' = f(x,y), \quad x \in [a,b] \, , \qquad\qquad (1.1)$$
$$y(a) = y_0,$$

where $y = [y^1(x), y^2(x), \ldots, y^s(x)]^T$ and y_0 is given.

All the numerical methods to be discussed here are known as discrete variable methods. That is, the methods will yield a sequence of approximations $y_n \approx y(x_n)$ on the set of points $x_{n+1} = x_n + h_n$, $n = 0, 1, 2, \ldots, N$, $x_0 = a$, $x_N = b$, where $h_n > 0$ is the stepsize. Most of the mathematical analysis presented here will treat the case where $h_n = h$, a constant. Later chapters, however, dealing with the implementation of methods will consider the case of variable stepsize. The mathematics in this case is usually more difficult, see for example Gear (1971) and more recently Gear and Tu (1974).

The basic assumption concerning (1.1) is that $f(x,y)$ satisfies a Lipschitz condition (with respect to the maximum norm),

$$\|f(x,y_1) - f(x,y_2)\| \leq L\| y_1 - y_2\| \qquad (1.2)$$

for all $x \in [a,b]$ and all s component vectors y_1 and y_2. It can then be shown that there exists a unique solution to the problem (1.1). The Lipschitz constant L plays a vital role in the theory and practice of numerical methods.

The purpose of this chapter is to briefly present in an informal fashion the various forms of discrete variable methods and to discuss sources of error. We shall introduce the methods for the scalar form of (1.1), s = 1, with respect to a constant stepsize h. Also the term "method" will be used in a loose sense, since at this stage we do not attempt to define precisely how a formula is implemented in practice.

2. DISCRETE VARIABLE METHODS

(i) Taylor Series Methods

Conceptually the simplest way of advancing the solution from the point x_n to x_{n+1} is by Taylor series (assuming appropriate differentiability),

$$y(x_{n+1}) = y(x_n) + h\Delta(x_n, y_n, h),$$
$$\text{where } \Delta(x,y,h) = y'(x) + \frac{h}{2}y''(x) + \frac{h}{3!}y'''(x) + \dots, \tag{2.1}$$

If this series is now truncated and $y(x_n)$ is replaced by the approximation y_n, we obtain with the aid of (1.1), the further approximation

$$y_{n+1} = y_n + h\,\phi\,(x_n, y_n, h), \; n = 0,1,2,\dots,$$
$$\text{where} \quad \phi(x,y,h) = f(x,y) + \frac{h}{2}f'(x,y) + \dots + \frac{h^{p-1}}{p!}\,f^{(p-1)}(x,y). \tag{2.2}$$

For p = 1 and 2 we have the formulas,

$$y_{n+1} = y_n + hf(x_n,y_n) \qquad \text{(Euler's method)}$$

$$y_{n+1} = y_n + h\left[f(x_n,y_n) + \frac{h}{2}(f_x(x_n,y_n) + f_y(x_n,y_n)\,f(x_n,y_n))\right],$$

from which, given y_0, approximations $\{y_n\}$ can be obtained, one step at a time. Such formulas are self starting and allow easy change of stepsize. Unfortunately their use in practice is severely restricted to those problems for which the higher total derivatives of f can be readily obtained. With recent developments in software, it has been possible in some cases to automatically generate algebraic expressions for f', f", ... It is therefore possible that the above Taylor series methods may be more widely used in the future, see Barton, Willers and Zahar (1971).

(ii) Runge-Kutta Methods.

Around the turn of the century Runge and subsequently Heun and Kutta introduced an idea which is equivalent to constructing a formula for $\emptyset$ which agrees with Δ as closely as possible without involving derivatives of f. This process of "matching" the Taylor series can be illustrated by setting

$$\emptyset(x,y,h) = c_1 f(x,y) + c_2 f(x + ha_2, y + b_{21} f(x,y)), \tag{2.3}$$

where c_1, c_2, a_2 and b_{21} are constants to be determined. Expanding both $\emptyset$ and Δ in powers of h, we obtain,

$$\emptyset(x,y,h) = (c_1+c_2) f(x,y) + hc_2\left[a_2 f_x(x,y) + b_{21} f_y(x,y)f(x,y)\right] + O(h^2),$$
$$\Delta(x,y,h) = f(x,y) + \tfrac{1}{2}h\left[f_x(x,y) + f_y(x,y)f(x,y)\right] + O(h^2),$$

yielding the equations,

$$c_1 + c_2 = 1, \quad c_2 a_2 = \tfrac{1}{2}, \quad c_2 b_{21} = \tfrac{1}{2}.$$

In general no agreement can be made which involves the $O(h^2)$ terms, and there results the family of solutions,

$$c_1 = 1 - \alpha, \quad c_2 = \alpha, \quad a_2 = b_{21} = \frac{1}{2\alpha},$$

where $\alpha \neq 0$ is a free parameter. When $\alpha = \frac{1}{2}$, we have the formula,

$$y_{n+1} = y_n + h\left[\tfrac{1}{2}f(x_n,y_n) + \tfrac{1}{2}f(x_n + h, y_n + hf(x_n,y_n))\right],$$

which was first obtained by Heun. This is an explicit formula which requires two evaluations of f per step.

The expression (2.3) and the above matching process can be extended to involve m evaluations of f, yielding the general m-stage explicit Runge-Kutta method,

$$
\begin{aligned}
y_{n+1} &= y_n + h\,\phi\,(x_n, y_n, h) \\
\phi(x,y,h) &= \sum_{r=1}^{m} c_r k_r, \quad k_1 = f(x,y), \\
k_r &= f(x + ha_r, y + h \sum_{s=1}^{r-1} b_{rs} k_s), \quad r=2,3,\ldots m.
\end{aligned}
\tag{2.4}
$$

The form of the method is ideal for practical computation; it is self starting and allows easy change of stepsize. The most well known formula is perhaps the 4-stage one,

$$
\begin{aligned}
y_{n+1} &= y_n + \frac{h}{6}(k_1 + 2k_2 + 2k_3 + k_4), \\
k_1 &= f(x_n, y_n), \quad k_2 = f(x_n + \tfrac{1}{2}h,\ y_n + \tfrac{1}{2}hk_1), \\
k_3 &= f(x_n + \tfrac{1}{2}h,\ y_n + \tfrac{1}{2}hk_2), \quad k_4 = f(x_n + h,\ y_n + hk_3).
\end{aligned}
\tag{2.4a}
$$

For larger values of m more accurate formulas can be constructed which may be used as the basis of highly efficient and widely used algorithms, see Chapter 5.

At the expense of the computational simplicity of the formulae (2.4), a broader class of formulae can be devised by allowing the coefficients $\{k_r\}$ to be defined as the solution of m equations (which in general are nonlinear), see Butcher (1964). The resulting so called m-stage implicit Runge-Kutta method is defined by

$$
\begin{aligned}
y_{n+1} &= y_n + h\phi\,(x_n, y_n, h), \\
\phi(x,y,h) &= \sum_{r=1}^{m} c_r k_r, \\
k_r &= f(x + ha_r,\ y + h \sum_{s=1}^{m} b_{rs} k_s), \quad r = 1,2,\ldots,m.
\end{aligned}
\tag{2.5}
$$

Since more parameters are available, such formulae are potentially more accurate than their corresponding m-stage explicit forms. This however, is at the cost of considerable computational complexity and at present the application of these methods appears to be restricted to a certain class of problems known as "stiff" equations, see Chapter 10.

It is possible to reduce the degree of implicitness in (2.5) by setting $b_{rs} = 0$ for $s > r$. As an example we have the 3-stage, semi-explicit method due to Butcher,

$$y_{n+1} = y_n + \frac{h}{6}(k_1 + 4k_2 + k_3)$$

$$k_1 = f(x_n, y_n),\ k_2 = f(x_n + \tfrac{1}{2}h,\ y_n + \tfrac{1}{4}hk_1 + \tfrac{1}{4}hk_2),$$

$$k_3 = f(x_n + h,\ y_n + hk_2).$$

One of the basic problems associated with the implementation of Runge-Kutta methods (and indeed all methods) is the choice of step-size h. For explicit methods a significant advance in this problem has been made through the introduction of "embedding forms", Chapter 3, a good account of which appears in Lapidus and Seinfeld (1971). For an m-stage process the idea is to use a sequence of additional formulae involving only the computed $k_1, k_2, \ldots, k_m$ which also yield approximations (of increasing accuracy) to $y(x_{n+1})$. From this information an estimate of the error in y_{n+1} can be obtained and the value of h may then be increased or decreased.

(iii) Linear Multistep Methods

When y_{n+1} is obtained by any of the above methods the approximation y_n is then discarded. Our approach in this section is to devise methods which make direct use of some previously computed values y_n, y_{n-1}, y_{n-2},... One way of constructing methods of this type is as follows. By integrating (1.1) we obtain the identity

$$y(x + \zeta) - y(x) = \int_x^{x+\zeta} f(t, y(t))\, dt, \qquad (2.6)$$

where x and $x + \zeta$ are any points in $[a,b]$. Now we replace $f(t,y(t))$ by an interpolating polynomial having the values $f_n = f(x_n, y_n)$ on a set of points x_n at which y_n has already been computed, or is just about to be computed. Let these interpolating points be $x_n, x_{n-1}, \ldots, x_{n-k}$, then the interpolating polynomial may be written, for example, in the Newton backward difference form,

$$P(t) = \sum_{r=0}^{k} (-)^r \binom{-s}{r} \nabla^r f_n, \qquad s = \frac{t-x_n}{h}.$$

This polynomial can now be integrated to yield from (2.6) various formulae which are determined by the position of x and $x+\zeta$ relative to the interpolating points. For example $x = x_n$, $\zeta = h$, leads to

$$y_{n+1} - y_n = \int_{x_n}^{x_{n+1}} P(t)dt = h \sum_{r=0}^{k} \gamma_r \nabla^r f_n \tag{2.7}$$

which is the well-known Adams-Bashforth method. Equivalently the formula can be expressed in Lagrangian form,

$$y_{n+1} - y_n = h \sum_{r=0}^{k} \beta_{kr} f_{n-r}. \tag{2.7a}$$

Similarly with $x=x_{n-1}$, $\zeta = h$, we obtain

$$y_n - y_{n-1} = h \sum_{r=0}^{k} \gamma_r^* \nabla^r f_n, \tag{2.8}$$

and in Lagrangian form,

$$y_n - y_{n-1} = h \sum_{r=0}^{k} \beta_{kr}^* f_{n-r}, \tag{2.8a}$$

which is the Adams-Moulton method. Some values of the constants γ_r, β_{kr}, γ_r^*, β_{kr}^* may be found in Henrici (1962, §5.1).

Many other formulae can be obtained in this fashion and similar formulae can be derived using other approaches, see Lambert (1973).

The general form of these methods is,

$$\sum_{j=0}^{k} \alpha_j y_{n+j} = h \sum_{j=0}^{k} \beta_j f_{n+j}, \quad n = 0,1,2,\ldots, \tag{2.9}$$

where α_j and β_j are constants, $\alpha_k \neq 0$, $|\alpha_0| + |\beta_0| \neq 0$; that is a linear relationship between y_{n+j}, f_{n+j}, $j = 0,1,2,\ldots,k$. This formula is called a <u>linear multistep method</u> or a linear k-step method. In order to generate the sequence of approximations $\{y_n\}$ it is first necessary to obtain k <u>starting values</u> $y_0, y_1, \ldots, y_{k-1}$. Then the computation takes one of two possible forms. Firstly, if $\beta_k = 0$, y_{n+k} is readily obtained and the method (2.9) is called an explicit multistep method. Secondly, if $\beta_k \neq 0$ the right hand side contains $f(x_{n+k}, y_{n+k})$ and in general it is now necessary to solve a nonlinear equation for y_{n+k}; (1.11) is then called an implicit multistep method. The Adams' formulas (2.7a), (2.8a) are examples of explicit and implicit methods respectively.

It would appear that an explicit multistep method is in general by far the simplest method computationally. In actual practice, however, explicit methods are very rarely used as methods in their own right. The reasons for this will be made precise in subsequent chapters, but as an illustration we can refer to the above Adams' formulas. For a given stepnumber k, the implicit Adams-Moulton method is potentially more accurate than the explicit Adams-Bashforth method. Further, when it comes to considering how the methods behave when the y_n values are subjected to small perturbations (which simulates the way in which the formulas behave when being implemented on a digital computer) the use of the explicit methods is restricted to significantly smaller values of stepsize h than that of the implicit methods, (this is on the basis of absolute stability - see Chapter 2.)

Accepting that implicit methods are generally preferred in practice we shall now indicate how they can be implemented; we shall refer to the form (2.9) but the same ideas apply to any other equivalent representation, for example (2.8). From (2.9),

$$y_{n+k} = \frac{h\beta_k}{\alpha_k} f(x_{n+k}, y_{n+k}) + g_n, \quad \beta_k \neq 0, \tag{2.10}$$

where g_n consists of known quantities y_{n+j}, f_{n+j}, $j=0, 1, \ldots, k-1$. It can be shown that (see Henrici (1962, §5.2) for precise conditions) if

$$h < \left|\frac{\alpha_k}{\beta_k}\right| \cdot \frac{1}{L},$$

then there exists a unique solution y_{n+k} of (2.10) which can be obtained by the iteration,

$$y_{n+k}^{(\nu+1)} = \frac{h\beta_k}{\alpha_k} f(x_{n+k}, y_{n+k}^{(\nu)}) + g_n, \quad \nu = 0, 1, 2, \ldots, \tag{2.11}$$

where $y_{n+k}^{(0)}$ is arbitrary. (It is also important to notice that the rate of convergence of (2.11) is governed by the size of $h\left|\frac{\beta_k}{\alpha_k}\right| L$). This iteration scheme is well suited to automatic computation and each step involves one evaluation of f. It is clearly to our advantage to choose $y_{n+k}^{(0)}$ as accurately as possible, a convenient choice being provided by the use of an explicit multistep formula. This explicit method is then called the predictor, and the implicit method (2.10) the corrector; the combined process is then called a predictor-corrector method. There are two approaches to the implementation of a predictor corrector method.

Firstly, the iteration (2.11) is applied until convergence (in practice until successive iterates differ by a sufficiently small amount). In this approach the number of f evaluations can vary from step to step and may be considerable; on the other hand, if we are prepared to reduce the size of h (and therefore increase the rate of convergence) convergence can always be obtained in less than a fixed number of iterations. This general process is called correcting to convergence.

In the second approach, at every step the corrector is applied a fixed number of times, say t (specified in advance); the resulting $y_{n+k}^{(t)}$ is then accepted as the approximation to $y(x_{n+k})$. This process is best described in the standard notation of Hull and Creemer (1963), where P denotes an application of the predictor, C a single application of the corrector and E an evaluation of f. To illustrate the various calculations, we shall use the Adams' pair,

$$\text{Predictor:} \quad y_{n+1} = y_n + \frac{h}{2}(3f_n - f_{n-1}),$$

$$\text{Corrector:} \quad y_{n+1} = y_n + \frac{h}{2}(f_{n+1} + f_n),$$

with the notation $f_n^{(\nu)} \equiv f(x_n, y_n^{(\nu)})$, we have for t = 1, the calculation,

$$P: y_{n+1}^{(0)} = y_n^{(1)} + \frac{h}{2}(3f_n^{(1)} - f_{n-1}^{(1)}),$$

$$\left.\begin{aligned} E: f_{n+1}^{(0)} &= f(x_{n+1}, y_{n+1}^{(0)}), \\ C: y_{n+1}^{(1)} &= y_n^{(1)} + \frac{h}{2}(f_n^{(1)} + f_{n+1}^{(0)}), \end{aligned}\right\} \tag{2.12}$$

$$E: f_{n+1}^{(1)} = f(x_{n+1}, y_{n+1}^{(1)}). \tag{2.12a}$$

This process is referred to as the PECE mode. A more economical calculation can be made by not performing the evaluation (2.12a) to give the so called PEC mode,

$$\left.\begin{aligned} P &: y_{n+1}^{(0)} = y_n^{(1)} + \frac{h}{2}(3f_n^{(0)} - f_{n-1}^{(0)}), \\ E &: f_{n+1}^{(0)} = f(x_{n+1}, y_{n+1}^{(0)}), \\ C &: y_{n+1}^{(1)} = y_n^{(1)} + \frac{h}{2}(f_n^{(0)} + f_{n+1}^{(0)}). \end{aligned}\right\} \tag{2.13}$$

Further, it is also possible to perform t corrections, that is, apply stages (2.12) and (2.13) t times, yielding the $P(EC)^tE$ and $P(EC)^t$ modes respectively.

One of the chief difficulties associated with the general use of multistep methods, as compared to the single-step Runge-Kutta methods, is in changing the stepsize h. On the other hand, for similar accuracy, multistep methods can be implemented in such a way that they usually require less evaluations of f per step than Runge-Kutta methods.

(iv) Block Methods

In the remaining parts of this section we shall very briefly present some lesser known methods. In most cases relatively little is known about their general numerical behaviour compared to that of conventional Runge-Kutta and multistep methods.

The first type of method is based on the idea of simultaneously producing a "block" of approximations $y_{n+1}, y_{n+2}, \ldots, y_{n+N}$; for example with N=2 we have the implicit block method due to Clippinger and Dimsdale,

$$\left.\begin{aligned} y_{n+1} - \frac{1}{2}y_{n+2} &= \frac{1}{2}y_n + \frac{h}{4}(f_n - f_{n+2}), \\ y_{n+2} &= y_n + \frac{h}{3}(f_n + 4f_{n+1} + f_{n+2}). \end{aligned}\right. \tag{2.14}$$

Hence given the previous block $(y_{n-1}, y_n)^T$, (2.14) is a formula which defines the next block $(y_{n+1}, y_{n+2})^T$; in this light it would perhaps be more appropriate to call it a one-step block implicit method - this is the approach developed by Shampine and Watts (1969, 1972), (it immediately suggests the notion of multistep block methods).

As shown by Butcher (1964), formula (2.14) can also be regarded as an implicit Runge-Kutta method with stepsize h = 2h. Block methods which are equivalent to explicit Runge-Kutta methods (defined with respect spect to step-size Nh) have been described by Rosser (1967), whose formulas offer the advantage of requiring less evaluations of f than the conventional Runge-Kutta methods. For examples and discussion also see Lambert (1973).

Implicit block methods as applied to systems of stiff differential equations has been developed in the spirit of linear multistep methods by Williams and de Hooge (1974).

(v) Hybrid Methods

Linear multistep methods use information only at the points $x_n=a+nh$, $n=0,1,2,\ldots$, whereas Runge-Kutta methods also use information at other points, for example $x_n + \frac{h}{2}$. A hybrid method attempts to combine these two features. Following Lambert (1973) this idea can be illustrated by defining the k-step hybrid formula

$$\sum_{j=0}^{k} \alpha_j y_{n+j} = h \sum_{j=0}^{k} \beta_j f_{n+j} + h\beta_\nu f_{n+\nu}, \tag{2.15}$$

where $\alpha_k \neq 0$, $|\alpha_0| + |\beta_0| \neq 0$, $\nu \neq 0,1,2\ldots,k$ and $f_{n+\nu} = f(x_{n+\nu}, y_{n+\nu})$. As it stands the formula is incomplete; to approximate $y_{n+\nu}$ a predictor formula is required. Through the choices ν and β_ν the form (2.15) offers the possibility of extending all the properties of conventional linear multistep methods. For a given k, one of the main objectives would be to extend the accuracy of (2.15) beyond that of the conventional form.

Hybrid methods will be further discussed in Chapter 3.

(vi) Multistep Multiderivative Methods

In this class of methods, linear multistep methods are extended to include higher derivatives of f(x,y). They can thus be regarded as the multistep generalisation of the Taylor Series methods. In the literature such methods are associated with the name of Obrechkoff. The k-step Obrechkoff method

$$\sum_{j=0}^{k} \alpha_j y_{n+j} = \sum_{i=1}^{\ell} h^i \left(\sum_{j=0}^{k} \beta_{ij} y_{n+j}^{(i)} \right),$$

uses the first ℓ derivatives of y. As an example with $k=1$, $\ell=2$, we have the implicit method,

$$y_{n+1} - y_n = \frac{h}{2}(y_{n+1}^{(1)} + y_n^{(1)}) - \frac{h^2}{12}(y_{n+1}^{(2)} - y_n^{(2)}),$$

whose potential accuracy is similar to that of the 4-stage explicit Runge-Kutta method (2.4a).

Naturally for efficient application of these methods it is essential that the total derivatives $y^{(i)}$ can be easily obtained. One such application area is stiff equations and efficient algorithms based on implicit Obrechkoff formulas have been devised by Liniger and Willoughby (1967), and Enright (1974); see Chapter 11 for further discussion.

3. APPLICATION TO SYSTEMS AND HIGHER ORDER EQUATIONS

In this section we will briefly show how the methods of §2 can be applied to both systems of first order equations and to systems of higher order equations.

For the first order system of s equations,

$$\begin{aligned} y' &= f(x,y), \; x \in [a,b], \\ y(a) &= y_0, \end{aligned} \tag{3.1}$$

where

$$y = [y^1(x),\, y^2(x),\ldots,y^s(x)]^T,\quad f(x,y) = [f^1(x,y),\, f^2(x,y),\ldots,f^s(x,y)]^T$$

advancing the solution from x_n to x_{n+1} by Taylor series can be applied component-wise. That is, for $i = 1,2,\ldots,s$,

$$y^i(x_{n+1}) = y^i(x_n) + h\frac{dy^i}{dx}(x_n) + \frac{h^2}{2}\frac{d^2y^i}{dx^2}(x_n) + \ldots \quad .$$

If, for example, we now truncate terms of order h^3 and higher, we obtain as in Section 2, the Taylor Series method,

$$y_{n+1}^i = y_n^i + hf_n^i + \frac{h^2}{2}\frac{d}{dx}f_n^i, \qquad i = 1,2,\ldots,s$$

in which y_n^i denotes the approximation to $y^i(x_n)$; and $f_n^i = f^i(x_n,y_n)$ where $y_n = [y_n^1, y_n^2,\ldots,y_n^s]^T$. Further, introducing the vector $f_n = [f_n^1, f_n^2,\ldots,f_n^s]^T$, the method can be written as

$$y_{n+1} = y_n + h(f_n + \frac{h}{2}f_n'), \tag{3.2}$$

where the vector f_n' is obtained from f_n by componentwise differentiation with respect to x. From

$$\frac{df^i}{dx} = \frac{\partial f^i}{\partial x}\frac{dx}{dx} + \sum_{j=1}^{s} \frac{\partial f^i}{\partial y^j}\frac{dy^j}{dx} \quad i = 1,2,\ldots,s,$$

we obtain in vector notation,

$$f'(x,y) = \frac{\partial f}{\partial x}(x,y) + J(x,y)\, f(x,y)$$

where $\partial f/\partial x$ is the vector obtained from f by component-wise partial differentiation and $J(x,y) = (\partial f^i/\partial y^j)$ is the $s \times s$ Jacobian matrix of the function f.

The Taylor method (3.2) is identical in form to the method for the scalar equation s=1; here scalars y and f have been replaced by s-component vectors. Similarly all the Runge-Kutta methods of §2 can be applied to systems by first replacing the scalars $y, f, \emptyset$ and k_r, $r=1,2,\ldots,m$, by corresponding s-component vectors.

One way of constructing linear multistep methods was by the so-called integration method which started with the identity (2.6). For systems of equations this can be applied component-wise,

$$y^i(x + \xi) - y^i(x) = \int_x^{x+\xi} f^i(t,y(t))\,dt, \quad i = 1,2,\ldots,s,$$

and the previous arguments follow through to yield linear multistep methods of the form (2.9) where now y_{n+j}, f_{n+j}, $j=0,1,\ldots,k$, are vectors with s-components. Similar remarks apply to all the other methods described in §2.

Consider now higher order equations. We will assume that a pth-order scalar equation (similar remarks apply to systems) is given in the form,

$$y^{(p)} = f(x,y,y',y'',\ldots,y^{(p-1)}), \quad x \in [a,b],$$

with initial conditions, (3.3)

$$y(a) = \eta_1, y'(a) = \eta_2, \ldots, y^{(p-1)}(a) = \eta_p,$$

where $y^{(i)} = d^i y/dx^i$. The most common technique for treating (3.3) is to transfer the equation into an equivalent first order system, by defining the new variables

$$y^i = y^{(i-1)}, \quad i = 1,2,\ldots,p.$$

From (3.3) we now obtain

$$(y^i)' = y^{i+1}, \quad i = 1,2,\ldots,p-1,$$
$$(y^p)' = f(x,y^1,y^2,\ldots,y^p), \tag{3.4}$$

with initial conditions

$$y^i(a) = \eta_i, \quad i = 1,2,\ldots,p.$$

This is a system of p first order equations and can be tackled by the previously discussed methods.

Alternatively, higher order equations can be treated directly and it is possible to devise Taylor methods, Runge-Kutta methods and multistep methods etc. which can be applied directly to the given equation (3.3), see, for example, Gear (1971); also see Chapter 6.

4. SOURCES OF ERROR

In this section we wish to give some indication of the nature of errors which can occur when implementing methods for ordinary differential equations. We shall consider the system (3.1) and the general class of methods defined in vector notation by

$$\sum_{i=0}^{k} \alpha_i y_{n+i} = h\phi_f(x_n\,;\, y_{n+k}, \ldots, y_n;\, h) \tag{4.1}$$

for $0 \leqslant n \leqslant N-k$, $b-a = Nh$ where $\{\alpha_i\}$ are constants, $\alpha_k \neq 0$, with given starting values $y_0, y_1, \ldots, y_{k-1}$. Here ϕ_f is a given vector-valued function which depends on f; if ϕ is independent of y_{n+k} the method is explicit, otherwise the method is implicit. This formulation includes many of the important methods discussed earlier.

Examples

1. The Two-Stage Explicit Runge-Kutta Method (Heun's Method)

$$k = 1, \quad \alpha_0 = -1, \quad \alpha_1 = 1,$$
$$\phi_f(t_0\,; u_1, u_0; h) \equiv \tfrac{1}{2}\left[f(t_0,u_0) + f(t_0 + h,\ u_0 + hf(t_0,\ u_0))\right].$$

2. General Linear Multistep Method

$$\phi_f(t_0;\ u_k,\ldots,u_0\,; h) \quad \equiv \quad \sum_{i=0}^{k} \beta_i f(t_0 + ih,\ u_i)$$

Consider now the problem (3.1) and suppose that the approximate solution $y_0, y_1, \ldots, y_N$, $Nh = b-a$ is given by (4.1).

Definition 4.1. The quantity $y_n - y(x_n)$ is called the global discretization error at $x = x_n$, $0 \leq n \leq N$.

One of the central problems of numerical methods for ordinary differential equations is the reliable control of global discretization error. A natural requirement is that this error can be made as small as we please by making h sufficiently small; this is the concept of convergence, see Chapter 2.

Definition 4.2. The local truncation error t_{n+k} of (4.1) at $x_{n+k} \in [a,b]$ is given by

$$t_{n+k} = \sum_{i=0}^{k} \alpha_i y(x_{n+i}) - h\phi_f(x_n;\, y(x_{n+k}),\ldots,y(x_n);\, h)\,. \tag{4.2}$$

The quantity t_{n+k} is the amount by which the true solution of (3.1) fails to satisfy (4.1) and may be regarded as a first measure of the accuracy of the formula. The significance of the term "local" can be explained as follows. If the method (4.1) is explicit and we assume $y_{n+r} = y(x_{n+r})$, $0 \leq r \leq k-1$, then from (4.1) and (4.2), there follows easily,

$$\|y_{n+k} - y(x_{n+k})\| = \frac{1}{|\alpha_k|}\,\|t_{n+k}\|\,,$$

and we have a simple and meaningful relationship between local truncation error and the error in y_{n+k}. Under the same assumption of "exact back values", a rather more complicated relationship between these two quantities can be shown for the case of an implicit method. Loosely, the local truncation error may be thought of as the error introduced by the formula at each step.

If we consider differential equations whose solutions y(x) are sufficiently differentiable then it is possible to obtain expressions for the local truncation error in terms of higher derivatives of y(x).

Examples

1. Consider Heun's method (and for notational convenience the scalar differential equation); omitting subscripts on x, the local truncation error is given by

$$y(x+h) - y(x) - \frac{h}{2}\left[f(x,y(x)) + f(x+h,\ y(x) + hf(x,y(x))\right].$$

Assuming $f(x,y)$ is sufficiently continuously differentiable with respect to both its variables, we obtain, using the Taylor expansion

$$f(x+h,\ y+q) = f(x,y) + f_x(x,y)h + f_y(x,y)q + O(h^2) + O(q^2).$$

Since $y'(x) = f(x,y)$, $y''(x) = f_x(x,y(x)) + f_y(x,y(x))y'(x)$, there follows $t_{n+1} = O(h^3)$. For Runge-Kutta methods the actual expressions involved in the leading terms of t_{n+1} are, in general, exceedingly complicated functions of higher mixed derivatives of f and are of little value in practice; for example in the above case the leading term is

$$\frac{h^3}{6}\{f_x f_y + ff_y^2 - \tfrac{1}{2}(f_{xx} + 2ff_{xy} + f^2 f_{yy})\}.$$

2. For the linear multistep method

$$\sum_{i=0}^{k} \alpha_i y_{n+i} = h \sum_{i=0}^{k} \beta_i f_{n+i}$$

assuming $y(x)$ is $q+1$ times continuously differentiable, we obtain using appropriate Taylor expansions,

$$\sum_{i=0}^{k} \alpha_i y(x+ih) - h \sum_{i=0}^{k} \beta_i y'(x+ih) \tag{4.3}$$
$$= C_0 y(x) + C_1 hy'(x) + C_2 h^2 y''(x) + \ldots + C_q h^q y^{(q)}(x) + O(h^{q+1}),$$

where the constants $C_0, C_1, \ldots, C_q$ are functions of the coefficients $\{\alpha_i\}$ and $\{\beta_i\}$. We find

$$C_0 = \alpha_0 + \alpha_1 + \ldots + \alpha_k,$$
$$C_1 = \alpha_1 + 2\alpha_2 + \ldots + k\alpha_k - (\beta_0 + \beta_1 + \ldots + \beta_k),$$

etc. (The expression (4.3) could be used as the starting point for deriving multistep formulas with a prescribed order of local truncation error). For methods used in practice the typical form for t_{n+k} is

$$t_{n+k} = C_{p+1}h^{p+1}y^{(p+1)}(x_n) + O(h^{p+2}), \quad p \geq 1,$$

and the size of the dominant term (the so-called principal local truncation error) can often be estimated and provide a means of controlling global accuracy. For the Adams-Bashforth 3-step method,

$$y_{n+3} = y_{n+2} + \frac{h}{12}(23f_{n+2} - 16f_{n+1} + 5f_n),$$

$$t_{n+3} = \frac{3}{8}h^4 y^{(4)}(x_n) + O(h^5).$$

It remains to be seen how the global discretization error is related to the local truncation error and the errors in the "back values" y_{n+i}, $0 \leq i \leq k-1$. Again for notational convenience consider the scalar case and let $e_n = y_n - y(x_n)$, $0 \leq n \leq N$, then from (4.1) and (4.2)

$$\begin{aligned}\sum_{i=0}^{k} \alpha_i e_{n+i} &= h[\phi_f(x_n; y_{n+k}, \ldots, y_n; h) \\ &\quad - \phi_f(x_n; y(x_{n+k}), \ldots, y(x_n); h)] - t_{n+k}, \\ &= h \sum_{i=0}^{k} \frac{\partial \phi_f}{\partial y_{n+i}}(x_n; \hat{y}_{n+k}, \ldots, \hat{y}_n; h)\, e_{n+i} - t_{n+k},\end{aligned} \tag{4.4}$$

where we have assumed that with respect to the variables y_{n+i}, $0 \leq i \leq k$, the Mean-Value theorem is applicable to the function ϕ_f. Hence beginning at n=0 with the "starting errors"

$$e_0, e_1, \ldots, e_{k-1},$$

e_k depends on a single local truncation error t_k plus propagated forms of the starting errors. In general e_{n+k} is determined by the local truncation error t_{n+k} and the way in which the difference equation (4.4) propagates the "back errors" e_n, $e_{n+1}, \ldots, e_{n+k-1}$. In practice a natural requirement is that local truncation errors are sufficiently small. In addition the method (as determined by the function ϕ_f and the coefficients $\{\alpha_i\}$) should be such that "propagated errors" do not grow too large (see Chapter 2 for more discussion).

In the above discussion no account has been taken of round-off error. In practice of course (using finite precision arithmetic) we obtain the computed approximations

$$\ldots, \tilde{y}_n, \tilde{y}_{n+1}, \ldots, \tilde{y}_{n+k}, \ldots,$$

which satisfy a perturbed form of (4.1),

$$\sum_{i=0}^{k} \alpha_i \tilde{y}_{n+i} = h\phi_f(x_n; \tilde{y}_{n+k}, \ldots, \tilde{y}_n; h) + R_{n+k},$$

where R_{n+k} contains the effects of round-off errors. There corresponds the global discretization error for the computed solution

$$\tilde{e}_n = \tilde{y}_n - y(x_n),$$

which satisfies a difference equation of the form (4.4) in which t_{n+k} is replaced by $R_{n+k} - t_{n+k}$; similar observations now apply since the round-off error acts in the same way as an additional local truncation error. Note, however, that $R_{n+k} - t_{n+k}$ cannot be made arbitrarily small as $h \to 0$.

Finally we wish to discuss another form of error which features in the theory and practice of recent algorithms. The difficulty associated with interpreting local truncation error is that we must assume exact "back values". Also when algorithms are tested it is extremely useful for comparison purposes to be able to actually compute the exact local truncation error; this essentially restricts the choice of test problems to those for which the true solutions are known. With this in mind an alternative form of error called the local error is more satisfactory, Hull et al (1972); also see Gear (1971) and Shampine and Gordon (1975).

Definition 4.3. The local error of (4.1) at $x_{n+k} \in [a,b]$ is given by

$$y_{n+k} - u(x_{n+k}, x_{n+k-1}),$$

where $u(x, x_{n+k-1})$ denotes the solution of the initial value problem

$$u' = f(x,u), \; x \in [x_{n+k-1}, b],$$
$$u(x_{n+k-1}, x_{n+k-1}) = y_{n+k-1} \quad . \tag{4.5}$$

It is now possible to make reliable comparisons between algorithms which employ estimates of local error, since at each step a test program could actually compute the local error by solving numerically (to sufficiently high accuracy) the initial value problem (4.5) over the range $[x_{n+k-1}, x_{n+k}]$. This can be done for systems of nonlinear equations whose true solutions are unknown.

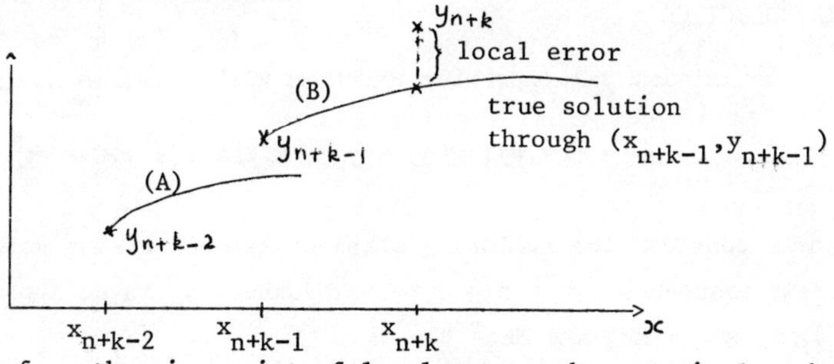

Further, from the view point of local error, the numerical method attempts to follow a particular solution curve, say curve (A), but due to some error, passes to a neighbouring solution curve (B). The behaviour of these solution curves therefore determines the character of the global discretization error.

2

CONVERGENCE AND STABILITY

1. INTRODUCTION

The initial value problem which we wish to solve numerically is

$$y' = f(x,y),\ y(a) = y_0;\quad x \in [a,b],\ y,f \in \mathbb{R}^s. \tag{1.1}$$

We shall consider the following class of general k-step method which generate sequences $(y_n \mid n = 0,1,\ldots,N)$ where y_n is an approximation to $y(x_n)$, $x_n = a + nh$, and $Nh = b - a$:

$$y_r = s_r(h),\quad 0 \leqslant r < k \quad \text{(starting values)}, \qquad \text{(i)}$$

$$\sum_{i=0}^{k} \alpha_i\, y_{n+i}/h = \phi_f(x_n;\, y_{n+k}, y_{n+k-1}, \ldots, y_n; h),\quad 0 \leqslant n \leqslant N-k, \qquad \text{(ii)} \tag{1.2}$$

It will be convenient to define the (first) characteristic polynomial $\rho(\theta)$ of (1.2) by

$$\rho(\theta) = \sum_{i=0}^{k} \alpha_i\, \theta^i.$$

The class (1.2) does not contain all discrete methods for solving (1.1) but it does contain a reasonably wide selection such as linear multistep methods (LMM), predictor-corrector methods (PC) (provided these are applied in mode $P(EC)^mE$ and not $P(EC)^m$), Runge-Kutta methods (RK) both explicit and implicit, and hybrid predictor-corrector methods (again only if applied in a mode which ends with an evaluation).

Examples

(a) LMM: $\phi_f := \sum_{i=0}^{k} \beta_i f(x_{n+i}, y_{n+i})$.

Here it is convenient also to define the second characteristic polynomial $\sigma(\theta)$ by

$$\sigma(\theta) := \sum_{i=0}^{k} \beta_i \theta^i.$$

(b) PECE: $\phi_f := \sum_{i=0}^{k-1} \beta_i f(x_{n+i}, y_{n+i})$

$$+ \beta_k f\left\{x_{n+k}, \frac{1}{\alpha_k^*} \sum_{i=0}^{k-1} [-\alpha_i^* y_{n+i} + h\beta_i^* f(x_{n+i}, y_{n+i})]\right\}.$$

Here ρ and σ will denote the characteristic polynomials of the corrector, ρ^* and σ^* those of the predictor.

(c) RK: $k=1$; $\phi_f := \sum_{r=1}^{m} c_r k_r$,

$$k_r = f\left\{x_n + ha_r, y_n + h \sum_{t=1}^{m} b_{rt} k_t\right\}, 1 \leq r \leq m.$$

2. CONVERGENCE

We recall that the global discretization error at x_n when (1.2) is applied to (1.1) is defined to be $y_n - y(x_n)$, $0 \leq n \leq N$. If this error tends to zero as h tends to zero, we say that the method is convergent, whence

Definition 1 A method of class (1.2) is convergent if, when applied to any problem of class (1.1),

$$\max_{0 \leq n \leq N} ||y_n - y(x_n)|| \to 0 \text{ as } h \to 0.$$

Note that the starting values given by (1.2(i)) as well as the

y_{n+k}, $0 \leq n \leq N-k$, must converge. It is important also to note that, since $Nh = b-a$, N will tend to infinity as $h \to 0$. We can bring out this point a little more clearly by considering what happens at a fixed station $x = a + nh$, $x \in [a,b]$, as $h \to 0$ (and therefore $n \to \infty$). Hence we can construct the following alternative (and equivalent) definition:

Definition 2 A method of class (1.2) is convergent if, when applied to any problem of class (1.1),

$$y_n \to y(x) \text{ as } h \to 0, \quad \text{where } n = (x-a)/h \,,$$

for any $x \in [a,b]$.

We now ask what condition a method of class (1.2) must satisfy in order that it be convergent. Intuitively one expects that a certain minimal level of local accuracy will be necessary. In discussing this it is convenient to work with the local discretization error, which is a differently normalized version of the local truncation error.

Definition 3 The local discretization error d_n of (1.2) at x_n is defined to be

$$d_r = y(x_r) - s_r(h), \quad 0 \leq r < k \,,$$

$$d_{n+k} = \frac{1}{\rho'(1)h}\left\{\sum_{i=0}^{k} \alpha_i \, y(x_{n+i}) - h\phi_f(x_n;\, y(x_{n+k}),\ldots,y(x_n);\, h)\right\}, \quad 0 \leq n \leq N-k\,, \qquad (2.1)$$

where $y(x)$ is the solution of (1.1). Thus d_n is seen to represent the amount by which the theoretical solution $y(x)$ fails to satisfy (1.2). The appropriate minimal level of local accuracy is now defined.

Definition 4 A method of class (1.2) is said to be consistent if

$$\max_{0 \leq n \leq N} ||d_n|| \to 0 \quad \text{as} \quad h \to 0.$$

It is consistent of order p if $\max_{0 \leq n \leq N} ||d_n|| = O(h^p)$.

Since $y(x) \in C^1[a,b]$, we may write the second of (2.1) in the form

$$d_{n+k} = \frac{1}{\rho'(1)h}\left\{\sum_{i=0}^{k}\alpha_i[y(x_n) + ihy'(x_n+\lambda_i ih)] - h\phi_f(x_n;y(x_{n+k}),\ldots,y(x_n);h)\right\}$$

$$= \frac{1}{\rho'(1)h}\left\{y(x_n)\sum_{i=0}^{k}\alpha_i + h\left[\sum_{i=0}^{k} i\alpha_i\, f(x_n+\lambda_i ih, y(x_n+\lambda_i ih))\right.\right.$$

$$\left.\left. - \phi_f(x_n;y(x_{n+k}),\ldots,y(x_n);\, h)\right]\right\},\quad 0 \leq n \leq N-k\,.$$

It follows that (1.2) is consistent iff

$$y_r \to y_0 \quad \text{as} \quad h \to 0, \quad 0 \leq r < k\,, \qquad \text{(i)}$$

$$\sum_{i=0}^{k} \alpha_i \;(\equiv \rho(1)) \;=\; 0\,, \qquad \text{(ii)} \qquad (2.2)$$

$$\phi_f(x_n;y(x_{n+k}),\ldots,y(x_n);h) \to \rho'(1)\, f(x_n,y(x_n))$$

$$\text{as} \quad h \to 0, \quad x_n = a+nh\,, \qquad \text{(iii)}$$

In the case of our examples, (2.2 (iii)) is satisfied iff

(a) LMM: $\sigma(1) = \rho'(1)$,

(b) PECE: $\sigma(1) = \rho'(1)$, $\rho^*(1) = 0$, (2.3)

(c) RK: $\sum_{r=1}^{m} c_r = \rho'(1)$.

The following heuristic argument illustrates the relationship of each of (2.2) to the concept of convergence. First, let us assume merely that y_n converges, in the sense of Definition 2, to <u>some</u> non-trivial function $Y(x)$. Then

$$y_{n+i} \to Y(x) \quad \text{as} \quad h \to 0, \quad n = (x-a)/h, \quad i = 0,1,\ldots,k,$$

and from (1.2 (ii)) we obtain

$$\sum_{i=0}^{k} \alpha_i \; Y(x) = 0 \; ,$$

whence (2.2 (ii)) holds. Note that this is so under the assumption that y_n tends to some function; no reference to the initial value problem (1.1) has been made. Further, since

$$\sum_{i=0}^{k} \alpha_i = 0 \; ,$$

we may write (1.2 (ii)) in the form

$$\sum_{i=0}^{k} i\alpha_i (y_{n+i} - y_n)/ih = \phi_f(x_n; y_{n+k}, \ldots, y_n; h).$$

If we assume that $Y(x) \in C^1[a,b]$, then

$$(y_{n+i} - y_n)/ih \to Y'(x) \text{ as } h \to 0,\; n = (x-a)/h,\; i = 0,1,\ldots,k,$$

and it follows that

$$\phi_f(x_n;\; Y(x_{n+k}),\ldots,Y(x_n);\; h) \to \rho'(1)\; Y'(x) \; .$$

The result (2.2 (iii)) now follows if we assume that $Y(x)$ does indeed satisfy the differential equation in (1.1); (2.2 (i)) follows if we also assume that $Y(x)$ satisfies the initial condition in (1.1). It is in this sense that consistency is the appropriate minimal level of local accuracy necessary for convergence. The following theorem can be formally established:

<u>Theorem 1</u> A convergent method of class (1.2) is necessarily consistent. (A proof, in the case of LMM, can be found in Henrici (1962); for wider classes of methods, which include (1.2), see Butcher (1966) and Spijker (1966).)

However, as we shall see in the next section, consistency is not sufficient for convergence of methods of class (1.2).

3. ZERO-STABILITY

We start by considering an example. The method

$$\begin{aligned} y_0 &= s_0(h), \quad y_1 = s_1(h), \\ y_{n+2} - 3y_{n+1} + 2y_n &= h(f_{n+1} - 2f_n), \quad 0 \leqslant n \leqslant N-2, \end{aligned} \tag{3.1}$$

applied to the problem

$$y' = 2x,\ y(0) = 0,\ (\text{solution } y(x) = x^2)$$

is readily seen, by (2.2) and (2.3 (a)), to be consistent if we choose starting values such that $s_0(h) \to 0$, $s_1(h) \to 0$ as $h \to 0$. The sequence (y_n) is then defined by

$$y_0 = s_0(h), \quad y_1 = s_1(h), \tag{i}$$

$$\begin{aligned} y_{n+2} - 3y_{n+1} + 2y_n &= 2h(x_{n+1} - 2x_n) \qquad (3.2) \\ &= 2h^2(1-n), \quad 0 \leqslant n \leqslant N-2. \quad \text{(ii)} \end{aligned}$$

The general solution of (3.2 (ii)) takes the form of Complementary Function + Particular Integral. The complementary function has the form $A\theta_1^n + B\theta_2^n$, where A and B are arbitrary constants and θ_1 (=1) and θ_2 (=2) are the roots of the characteristic polynomial of the difference equation,

$$\theta^2 - 3\theta + 2\ ,$$

which is, of course, $\rho(\theta)$. The particular integral is readily found to be $n(n-1)h^2$, whence the general solution of (3.2 (ii)) is

$$y_n = A + B.2^n + n(n-1)\ h^2\ ,$$

and the particular solution satisfying (3.2 (i)) is

$$y_n = [2s_0(h) - s_1(h)] + [s_1(h) - s_0(h)]\ 2^n + n(n-1)h^2. \tag{3.3}$$

There exists a particular choice, $s_0(h) = 0$, $s_1(h) = 0$, (which does not, incidentally, coincide with the exact starting values $y(0) = 0$, $y(h) = h^2$) such that

$$y_n = n(n-1)h^2 = (nh)^2 - (nh)h$$
$$\to x^2 \quad \text{as} \quad h \to 0, \quad n = x/h.$$

If, however, $s_0(h) \to 0$, $s_1(h) \to 0$ as $h \to 0$ such that $s_1(h) - s_0(h) = O(h^q)$, $q > 0$, then, since

$$\lim_{\substack{h\to 0\\ nh=x}} h^q\theta^n = x^q \lim_{n\to\infty} \theta^n/n^q = \infty \quad \text{if} \quad |\theta| > 1, \tag{3.4}$$

the second term of (3.3) tends to infinity, and the method is not convergent.

A similar situation arises if we revert to the special starting values $s_0(h) = 0 = s_1(h)$ but perturb the difference equation (3.2 (ii)) by a small amount $\delta(h) = O(h^q)$, $q > 0$ to obtain

$$y_{n+2} - 3y_{n+1} + 2y_n = 2h^2(1-n) + \delta(h) ,$$

whose general solution is

$$y_n = A + B.2^n + n(n-1)h^2 - n\,\delta(h).$$

Fitting the starting values $s_0(h) = 0 = s_1(h)$ yields

$$y_n = -\delta(h) + \delta(h)2^n + n(n-1)h^2 - n\,\delta(h),$$

and again there is no convergence. Thus, even if we had a means of finding, in general, the special starting values which annihilate the divergent component in theory, in practice, any subsequent perturbation of the difference equation (such as rounding error) will cause the divergent component to reappear.

Thus the method (3.1) is sensitive to small perturbations both in its starting values and in the difference equation; it is <u>unstable</u> in the sense that the resulting perturbations in the solution y_n are unbounded, <u>even in the limit as $h \to 0$, $nh=x$</u>. (The underlined phrase motivates us to call the stability property we shall presently define <u>zero</u>-stability.)

Before formulating a definition of zero-stability, it is instructive to look at an analogous situation concerning the initial value problem (1.1) itself. Consider the class of perturbed initial value problems

$$z' = f(x,z) + \delta(x),\ z(a) = y_0 + \delta,\quad x \in [a,b]\ ,$$

where $(\delta(x),\ \delta)$ is the perturbation and $z(x)$ the resulting perturbed solution.

Definition 5 (Hahn (1967), Stetter (1971).) Let $(\delta(x),\ \delta)$, $(\delta^*(x), \delta^*)$ be any two perturbations and let $z(x)$, $z^*(x)$ be the resulting perturbed solutions. Then if there exists a positive constant S such that, for all $x \in [a,b]$,

$$||z(x) - z^*(x)|| \leqslant S\varepsilon$$

whenever $$||\delta(x) - \delta^*(x)|| \leqslant \varepsilon \text{ and } ||\delta - \delta^*|| \leqslant \varepsilon\ ,$$

we say that the initial value problem (1.1) is totally stable.

Total stability is not much to demand of an initial value problem. (Indeed, the Lipschitz condition satisfied by $f(x,y)$, which is assumed throughout this series of lectures, is sufficient to imply total stability; see Gear (1971).) It is intuitively clear that if an initial value problem is not totally stable, then we have no chance of obtaining an acceptable numerical solution by any discretization method - nor have we unless the method itself satisfies an analogous stability property, which we now define.

Consider the class of perturbations of the method (1.2)

$$z_r = s_r(h) + \delta_r, \quad 0 \leqslant r < k\ ,$$

$$\sum_{i=0}^{k} \alpha_i\ z_{n+i}/h = \phi_f(x_n,\ z_{n+k},\ldots,z_n,h) + \delta_{n+k}, \qquad (3.5)$$

$$0 \leqslant n \leqslant N-k,$$

where $(\delta_n | n = 0,1,\ldots,N)$ is the perturbation and $(z_n | n = 0,1,\ldots,N)$ the resulting perturbed solution.

Definition 6 Let $(\delta_n | n = 0,1,\ldots,N)$, $(\delta_n^* | n = 0,1,\ldots,N)$ be any two perturbations and let $(z_n | n = 0,1,\ldots,N)$, $(z_n^* | n = 0,1,\ldots,N)$ be the resulting perturbed solutions. Then if there exist constants h_0 and S such that for all $h \in (0,h_0]$,

$$||z_n - z_n^*|| \leq S\varepsilon, \quad \text{for} \quad 0 \leq n \leq N$$

whenever $$||\delta_n - \delta_n^*|| \leq \varepsilon, \quad \text{for} \quad 0 \leq n \leq N, \tag{3.6}$$

we say that the method (1.2) is <u>zero-stable</u>.

(This property is frequently also called <u>stability</u> or <u>D-stability</u>, the latter referring to the original paper by Dahlquist (1956).)

We have already demonstrated that the method (3.1) of our example is not zero-stable. Looking back at that example, it is clear that it is the existence of the root +2 of $\rho(\theta)$ which precludes the possibility of zero-stability. This result generalizes easily. Consider the initial-value problem

$$y' = 0, \; y(0) = 0,$$

and assume - as is always the case - that

$$\phi_{f\equiv 0}(x_n; y_{n+k}, \ldots, y_n; h) \equiv 0,$$

so that (3.5) becomes

$$\begin{aligned} z_r &= s_r(h) + \delta_r, \quad 0 \leq r < k, \\ \sum_{i=0}^{k} \alpha_i z_{n+i} &= 0, \quad 0 \leq n \leq N-k. \end{aligned} \tag{3.7}$$

Suppose that $\rho(\theta)$ has a root ψ such that $|\psi| > 1$. Consider the following pair of perturbations:

$$\begin{aligned} (\delta_n) : \quad \delta_r &= \psi^r \omega - s_r(h), \quad 0 \leq r < k, \\ \delta_{n+k} &= 0, \quad 0 \leq n \leq N-k, \\ (\delta_n^*) : \quad \delta_r^* &= - s_r(h), \quad 0 \leq r < k, \\ \delta_{n+k}^* &= 0, \quad 0 \leq n \leq N-k. \end{aligned}$$

Since $$\sum_{i=0}^{k} \alpha_i \psi^{n+i} = \psi^n \rho(\psi) = 0,$$

it follows from (3.7) that the corresponding perturbed solutions are

$$z_n = \psi^n \omega, \; z_n^* = 0, \quad 0 \leq n \leq N .$$

Thus,
$$\| z_n - z_n^* \| = \| \psi^n \omega \| = |\psi|^n \, \|\omega\| \, ,$$

while
$$\| \delta_n - \delta_n^* \| \leq \max_{0 \leq r < k} \| \psi^r \omega \| =: \varepsilon$$

and since $N \to \infty$ as $h \to 0$, it is clear that (3.6) cannot hold.

Suppose now that $\rho(\theta)$ has a root ψ of multiplicity greater than 1 such that $|\psi| = 1$, and consider the perturbations

$$\begin{aligned} (\delta_n) \quad : \quad & \delta_r = r\psi^r \omega - s_r(h) && 0 \leq r < k \\ & \delta_{n+k} = 0 && 0 \leq n \leq N-k \\ (\delta_n^*) \quad & \delta_r^* = - s_r(h) && 0 \leq r < k \\ & \delta_{n+k}^* = 0 && 0 \leq n \leq N-k \end{aligned}$$

Since $\sum_{i=0}^{k} \alpha_i (n+i) \psi^{n+i} = \psi^n \, [\, n\rho(\psi) + \rho'(\psi)] = 0$, it follows from (3.7) that the corresponding perturbed solutions are

$$z_n = n\psi^n \omega, \; z_n^* = 0, \quad 0 \leq n \leq N .$$

Thus
$$\| z_n - z_n^* \| = \| n\psi^n \omega \| = n \|\omega\| ,$$

while
$$\| \delta_n - \delta_n^* \| \leq \max_{0 \leq r < k} \| r\psi^r \omega \| =: \varepsilon$$

and once again (3.6) cannot hold.

Thus, if $\rho(\theta)$ has a root outside the unit circle, or a multiple root on the unit circle, the method (1.2) cannot be zero-stable.

Definition 7 The method (1.2) is said to satisfy the root condition if the roots of the characteristic polynomial $\rho(\theta)$ all lie within or on the unit circle, those on the unit circle being simple.

Note that if the method (1.2) is consistent, then by (2.2 (ii)) $\rho(\theta)$ necessarily has a root $\theta_1 = +1$. It is convenient to make a further definition at this stage:

Definition 8 The method (1.2) is said to satisfy the strong root condition if its characteristic polynomal $\rho(\theta)$ has a simple root at +1, all the remaining roots lying strictly within the unit circle.

The roots θ_ν, $\nu = 1,2,\ldots,k$, of $\rho(\theta)$ for a consistent method satisfying the root condition are frequently categorized in the following ways:

$\theta_1 = +1$: principal root,

$\theta_\nu, |\theta_\nu| \leq 1, \nu = 2,3,\ldots,k$: $\left\{\begin{array}{l}\text{spurious}\\ \text{extraneous}\end{array}\right\}$ roots

$\theta_\nu, |\theta_\nu| = 1, \nu = 1,2,\ldots,m$: essential roots

$\theta_\nu, |\theta_\nu| < 1, \nu = m+1,\ldots,k$: non-essential roots.

We have already shown that for a method of class (1.2) zero-stability implies satisfaction of the root condition. The converse is also true, and the following important result can be established (see, for example, Isaacson and Keller (1966)):

Theorem 2 A method of class (1.2) is zero-stable if and only if it satisfies the root condition.

We can now state the fundamental theorem concerning convergence:

Theorem 3 A method of class (1.2) is convergent if and only if it is both consistent and zero-stable.

Note that a consistent one-step method of class (1.2) necessarily satisfies the root condition (and is therefore zero-stable) since, by (2.2 (ii)), the only root of $\rho(\theta)$ is $\theta_1 = +1$. Hence

Theorem 4 A one-step method of class (1.2) is convergent if and only if it is consistent.

Theorem 3 has occupied a central position in the development of the numerical analysis of differential equations. It was first proved, in the case of LMM, by Dahlquist (1956) - see also Henrici (1962). A proof for the class (1.2) may be found in Isaacson and

Keller (1966). For yet wider classes of methods see Gear (1965), Butcher (1966), Spijker (1966), Chartres and Stepleman (1972) and Mäkelä, Nevanlinna and Sipilä (1974); see also Watt (1967) and Stetter (1973).

[The proof of Theorem 3 depends centrally on the assumption that $f(x,y)$ in (1.1) satisfies a Lipschitz condition w.r.t. y. Recently, Taubert (1974) has shown that if (1.1) has a unique solution but $f(x,y)$ is only continuous w.r.t. y, then Theorem 3 no longer holds, and in the case of LMM, the necessary and sufficient condition for convergence is that the method be consistent and satisfy the strong root condition.]

Qualitatively speaking, the results of this section can be summarized by saying that consistency controls the magnitude of the local discretization error, while zero-stability controls the manner in which this and other errors are propagated in the limit as $h \to 0$, $Nh = b-a$; both properties are essential if convergence is to be achieved. Convergence is a minimal property to expect of a numerical method (just as total stability is a minimal property to expect of an initial-value problem). There is no practical use to which we can put methods which are not convergent. On the other hand, as we shall see later, it is by no means true that all convergent methods are suitable for practical computation.

4. ATTAINABLE ORDER OF ZERO-STABLE LMM

The LMM

$$\sum_{i=0}^{k} \alpha_i \, y_{n+i} = h \sum_{i=0}^{k} \beta_i \, f(x_{n+i}, y_{n+i}) \tag{4.1}$$

has essentially $2k+1$ free parameters if it is implicit, and $2k$ if it is explicit (for one parameter is lost upon normalization). On the other hand, it is easily established that if (4.1) is to have order p, then the free parameters must satisfy $p+1$ linear conditions. Thus the highest order we can expect (4.1) to achieve is $2k$ (implicit) or $2k-1$ (explicit). However, these maximal orders cannot be achieved without violating the root condition:

<u>Theorem 5</u> No zero-stable LMM of stepnumber k can have order

exceeding k+1 when k is odd or k+2 when k is even.

A proof can be found in Henrici (1962).

<u>Definition 9</u> A zero-stable LMM of stepnumber k is said to be <u>optimal</u> if it has order k+2.

It can be shown that for optimal LMM all the roots of $\rho(\theta)$ are necessarily essential roots.

Theorem 5, which it must be stressed, applies only to LMM and not to the general class (1.2), was the motivation behind the development of modified LMM, such as the hybrid methods of Gragg and Stetter (1964), Gear (1964), and Butcher (1965).

5. STABILITY FOR FIXED STEP-LENGTH

In section 2 we established necessary and sufficient conditions for a method of class (1.2) to be convergent. Throughout the remaining sections of this chapter it will be assumed that all methods of class (1.2) satisfy these conditions. That convergence is not in itself a guarantee that a method will yield acceptable numerical results is illustrated by the following example. We wish to solve the problem

$$y' = -10(y-1)^2,\ y(0) = 2,\ (\text{solution } y(x) = 1 + 1/(1+10x)), \tag{5.1}$$

using the following explicit hybrid algorithm of Gragg and Stetter (1964) in $P_\nu EP_H E$ mode:

$$y_0 = 2,\quad y_1 = 1 + 1/(1+10h),$$

$$P_\nu:\ y_{n+2.7} - 2.267025y_{n+2} - 3.5721y_{n+1} + 4.839125y_n$$

$$= h(f_{n+2} - 4.6751f_{n+1} - 1.73105f_n), \tag{5.2}$$

$$P_H:\ y_{n+3} - y_{n+2} = \frac{h}{714}(221f_{n+2} - 7f_{n+1} + 500f_{n+2.7}).$$

The characteristic polynomial of (5.2) is $\rho(\theta) = \theta^3 - \theta^2$, and the method is clearly zero-stable; it is also consistent, and indeed has order four. The numerical results of Table 1 are obtained with steplengths h = 0.1 and h = 0.01,

x_n	$y_n - y(x_n)$	
	h = 0.1	h = 0.01
0.03	-	-567×10^{-8}
0.04	-	-723×10^{-8}
0.05	-	-491×10^{-8}
0.10	-	-379×10^{-8}
0.20	-	-186×10^{-8}
0.30	-0.118	-106×10^{-8}
0.40	-0.072	-68×10^{-8}
0.50	-0.287	-48×10^{-8}
0.60	-0.361	-34×10^{-8}
0.70	-1.929	-27×10^{-8}
0.80	-40.131	-21×10^{-8}
0.90	-2×10^{6}	-17×10^{-8}

Table 1

What is significant about these results is that when $h = 0.01$ the global error decreases as n increases, whereas when $h = 0.1$ it increases, quite disastrously, as n increases. The apparent failure of the method when $h = 0.1$ is a serious one, since an estimate of the local discretization error when $h = 0.1$ indicates an acceptable level of local accuracy. What is happening is that small errors (both discretization and rounding) are being propagated in a stable manner when $h = 0.01$, but in an unstable manner when $h = 0.1$. It is clearly imperative that we can make some estimate of the maximum steplength for which stable propagation of error will occur when (5.2) is applied to (5.1).

Zero-stability ensures the stable propagation of error, in the limit as $h \to 0$. The above numerical results suggest that, for the method (5.2) applied to the problem (5.1), there does exist a positive h_0 such that for $h \in (0,h_0)$ stable propagation of error will occur. This is not always the case. There exist convergent methods of class (1.2), such as Simpson's rule,

$$y_{n+2} - y_n = \frac{h}{3}(f_{n+2} + 4f_{n+1} + f_n), \tag{5.3}$$

for which the error will be propagated in an unstable manner for all positive h, no matter how small. This is not incompatible with convergence. If we used (5.3) to solve (5.1), we should find that as we successively reduced the steplength, then the station x^* at which the global error first exceeds a fixed given bound moves progressively to the right. In the limit as $h \to 0$, convergence over any finite interval $[a,b]$ is achieved, yet for any fixed positive h we could find a finite x^* at which the global error exceeds the given bound.

It is clear that we require a new stability definition in which we regard the steplength as being fixed and demand that the error be propagated in a stable manner as $n \to \infty$.

6. REGIONS OF ABSOLUTE STABILITY

Ideally, the definition we seek should be something like the following:

> "A method of class (1.2) will be said to be absolutely stable for a given fixed steplength and for a given initial value problem (1.1) if the global error $\varepsilon_n := y_n - y(x_n)$ remains bounded as $n \to \infty$."

The snag with this tentative definition is the dependence on the initial value problem (1.1). Unless we can find a definition which is independent of the particular initial value problem, it will be impossible to compare the absolute stability properties of different methods of class (1.2). Accordingly, we settle on a specific initial value problem, and consider the test equation

$$y' = \lambda y, \tag{6.1}$$

where λ is a complex constant. The reason for choosing λ to be complex is the following. If we assume that f is differentiable with respect to y, then the local behaviour of the general initial value problem (1.1) is determined, to a first approximation, by the solution of the linearised equation

$$y' = \frac{\partial f}{\partial y} y ,$$

where $\partial f/\partial y$ is the Jacobian matrix; we model this linearised equation by

$$y' = Ay, \tag{6.2}$$

where A is a constant $s \times s$ matrix. If we assume that A has s distinct eigenvalues λ_t, $t = 0,1,\ldots,s$, then there exists a non-singular matrix H such that

$$H^{-1} AH = \Lambda = \text{diag}(\lambda_1, \lambda_2, \ldots, \lambda_s).$$

If we apply the transformation

$$y = Hz,$$

(6.2) becomes

$$z' = H^{-1} AHz = \Lambda z,$$

which is an uncoupled system of equations of the form of (6.2). For all commonly used methods of class (1.2) the diagonalizing transformation $y_n = Hz_n$ can be applied to (1.2) to yield a difference initial value problem for z_n; the stability properties of z_n are clearly identical with those of y_n, so that (6.1) is an acceptable model, if we interpret λ as representing, locally, <u>any</u> of the eigenvalues of the Jacobian, which may, of course, be complex. (If $s = 1$, it is sufficient to consider real λ.)

We now make the assumption - which is satisfied by all commonly used methods of class (1.2) - that when we substitute the test equation (6.1) for $y' = f(x,y)$ in (1.1), then ϕ_f in (1.2) becomes linear in y_{n+i}, $i = 0,1,\ldots,k$. Specifically, we assume that

$$\phi_{f=\lambda y}(x_n; y_{n+k}, \ldots, y_n; h) = \lambda \sum_{i=0}^{k} \gamma_i(h\lambda) y_{n+i}. \tag{6.3}$$

Examples (a) LMM: $\gamma_i(h\lambda) = \beta_i$, $i = 0,1,\ldots,k$,

(b) PECE: $\gamma_i(h\gamma) = \beta_i - \frac{\beta k}{\alpha_k^*} (\alpha_i^* - \beta_i^* h\lambda)$,

($i = 0,1,\ldots,k$ (if we formally define $\beta_k^* = 0$)).

(c) RK: $\gamma_1(h\lambda) \equiv 0$; $\gamma_0(h\lambda) = P_{m-1}(h\lambda)/Q_m(h\lambda)$,

where P_{m-1} is a polynomial of degree at most $m-1$, Q_m is a polynomial of degree at most m, $Q_m(0) = 1$; if the method is explicit, Q_m has degree zero.

Consider the application of (1.2) to the initial value problem

$$y' = \lambda y, \; y(a) = y_0,$$

for complex λ.

Then, from the definition of the local discretization error d_n of (1.2) given in equation (2.1),

$$y(x_r) = s_r(h) + d_r, \qquad 0 \leq r < k,$$

$$\sum_{i=0}^{k} \alpha_i y(x_{n+i}) = h\phi_{f=\lambda y}(x_n; y(x_{n+k}), \ldots, y(x_n); h) + h\rho'(1) d_{n+k}$$

$$= h\lambda \sum_{i=0}^{k} \gamma_i(h\lambda) y(x_{n+i}) + h\rho'(1) d_{n+k}, \qquad 0 \leq n,$$

by (6.3). But the numerical values y_n produced by (1.2) satisfy

$$y_r = s_r(h) + \tau_r, \qquad 0 \leq r < k,$$

$$\sum_{i=0}^{k} \alpha_i y_{n+i} = h\lambda \sum_{i=0}^{k} \gamma_i(h\lambda) y_{n+i} + \tau_{n+k}, \qquad 0 \leq n,$$

where τ_n represents the local rounding error. On subtracting, we have that the global error $\varepsilon_n := y_n - y(x_n)$ satisfies

$$\varepsilon_r = \tau_r - d_r, \qquad 0 \leq r < k,$$

$$\sum_{i=0}^{k} [\alpha_i - h\lambda\gamma_i(h\lambda)]\, \varepsilon_{n+i} = \tau_{n+k} - h\rho'(1) d_{n+k}, \qquad 0 \leq n, \tag{6.4}$$

whose solution is of the form Complementary Function + Particular Integral.

Define the stability polynomial of (1.2) to be

$$\pi(r;\ h\lambda) := \sum_{i=0}^{k} [\alpha_i - h\lambda\gamma_i(h\lambda)]r^i \tag{6.5}$$

and let its roots be r_ν, $\nu = 0,1,\ldots,k$. The complementary function for (6.4) takes the form

$$\sum_{\nu=1}^{k} A_\nu\ r_\nu^n \tag{6.6}$$

in the case when the r_ν are distinct. (If, for example, r_ν is a double root, its contribution to (6.6) is $(A_\nu + nB_\nu)r^n$.) The particular integral for (6.4) can be kept small if the local discretization and rounding errors are kept sufficiently small, and it is clear that the global error will be propagated in a stable manner as $n \to \infty$, h fixed, if the roots of the stability polynomial (6.5) satisfy

$$|r_\nu| < 1, \quad \nu = 1,2,\ldots\ .$$

Definition 10 The method (1.2) is said to be absolutely stable for a given $h\lambda$ if, for that $h\lambda$, all the roots of the stability polynomial lie within the unit circle. A region $\mathcal{R}_A$ of the complex plane is said to be a region of absolute stability of (1.2) if (1.2) is absolutely stable for all $h\lambda \in \mathcal{R}_A$. Note that in the case of a scalar initial value problem - or of a system for which it is known that the eigenvalues of the Jacobian are always real - it is appropriate to talk of an interval of absolute stability, which is simply the intersection of $\mathcal{R}_A$ with the real axis.

For our examples, the stability polynomial takes the following forms:

(a) LMM: $\pi(r;\ h\lambda) = \rho(r) - h\lambda\sigma(r)$,

(b) PECE: $\pi(r;\ h\lambda) = \rho(r) - h\lambda\sigma(r) + h\lambda\,\dfrac{\beta_k}{\alpha_k^*}\,[\rho^*(r) - h\lambda\sigma^*(r)]$,

(c) RK: $\pi(r;\ h\lambda) = \alpha_1 r + \alpha_0 - h\lambda\gamma_0(h\lambda)$, where γ_0 is a polynomial in $h\lambda$ if the method is explicit, and a rational function if it is implicit.

The form of the stability polynomial for more general PC algorithms and for hybrid algorithms are given by Lambert (1973)

The roots of a polynomial are continuous functions of its

coefficients. Hence, by (6.5) the roots r_ν, $\nu = 0,1,\ldots,k$ of the stability polynomial $\pi(r;\,h\lambda)$ tend to the roots θ_ν, $\nu = 0,1,\ldots,k$ of the characteristic polynomial $\rho(\theta)$ as $h \to 0$. In particular,

$$r_1 \to \theta_1 = +1 \quad \text{as} \quad h \to 0\,. \tag{6.7}$$

We can find a stronger result concerning the convergence of r_1 to θ_1. If the method (1.2) has order p, then

$$\max_{k\leqslant n\leqslant N}\left\| \frac{1}{\alpha_k}\left\{\sum_{i=0}^{k}\alpha_i y(x_{n+i}) - h\phi_f(x_n,y(x_{n+k}),\ldots,y(x_n),h)\right\}\right\| = O(h^{p+1})$$

Setting $f = \lambda y$, $y(x_n) = y_0 e^{\lambda(x_n-a)}$, we obtain

$$\max_{k\leqslant n\leqslant N}\left\| \sum_{i=0}^{k}\alpha_i y_0 e^{\lambda(x_n+ih-a)} - h\lambda\sum_{i=0}^{k}\gamma_i(h\lambda)e^{\lambda(x_n+ih-a)}\right\| = O(h^{p+1}),$$

or
$$\left|\sum_{i=0}^{k}[\alpha_i - h\lambda\gamma_i(h\lambda)]e^{ih\lambda}\right| \max_{k\leqslant n\leqslant N}\left\|e^{\lambda(x_n-a)}y_0\right\| = O(h^{p+1}),$$

whence
$$\pi(e^{h\lambda};\,h\lambda) = O(h^{p+1}).$$

But $\pi(e^{h\lambda};\,h\lambda) = \text{const.}\,(e^{h\lambda}-r_1)\,(e^{h\lambda}-r_2)\,\ldots\,(e^{h\lambda}-r_k)$, and since, by zero-stability, r_1 is the only one of the r_ν which can tend to +1 as $h \to 0$, it follows that $e^{h\lambda}-r_1 = O(h^{p+1})$, or

$$r_1 = e^{h\lambda} + O(h^{p+1}), \tag{6.8}$$

which is the required stronger form of (6.7). It follows from (6.8) that when $h\lambda = 0$, $|r_1| = 1$, and when Re $h\lambda$ is small and positive, $|r_1| > 1$. It follows that the boundary $\partial\mathcal{R}_A$ of the region $\mathcal{R}_A$ of absolute stability always passes through the origin in the complex $h\lambda$-plane, and that the interior of $\mathcal{R}_A$ always lies to the left of the origin.

In figure 1 are given sketches of $\mathcal{R}_A$ for three examples:

(i) Euler's method, $y_{n+1} - y_n = hf(x_n, y_n)$,

(ii) Backward Euler method, $y_{n+1} - y_n = hf(x_{n+1}, y_{n+1})$,

(iii) Any 4-th order 4-stage explicit RK method.

(It can be shown (Lambert, 1973) that for m = 1,2,3,4, all m-stage explicit RK methods of order m have the same region of absolute stability.)

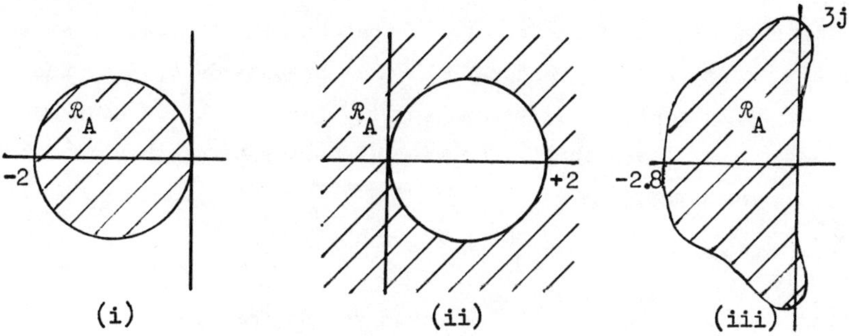

Figure 1

The most usual method for finding the region of absolute stability is the so-called boundary locus method. A point $h\lambda$ will lie on $\partial\mathcal{R}_A$ if, for that $h\lambda$, the stability polynomial has a root of modulus one, that is, if there exists real ψ such that

$$\pi(e^{j\psi};\ h\lambda) = 0, \qquad j^2 = -1. \tag{6.9}$$

This equation is solved numerically for a range of ψ, typically $\psi = 0^\circ, 30^\circ, 60^\circ, \ldots$, and a curve fitted through the resulting points in the complex $h\lambda$-plane to give an approximation to $\partial\mathcal{R}_A$; for simple methods it is frequently possible to solve (6.9) analytically. Other methods for finding $\mathcal{R}_A$, depending on the Schur criterion and the Routh-Hurwitz criterion are discussed by Lambert (1973). The paper by Miller (1971) is also useful in this context.

It is possible to use the region of absolute stability to monitor the steplength in the following fashion. From time to time as the computation proceeds, the eigenvalues of the Jacobian are computed approximately, and h is readjusted so that the product of h with any eigenvalue lies within $\mathcal{R}_A$. Such a strategy is viable for a scalar initial value problem - or perhaps for a small system or a system with special structure - but it is tedious to implement for a large general system. Many modern algorithms, such as that of Gear (1971), do not test for absolute stability. Nevertheless, the size and shape of $\mathcal{R}_A$ are important in assessing the value of a particular method and in comparing it with other methods.

7. STRONG STABILITY

We have already seen from (6.8) that when $h\lambda$ is real, positive and small, then $|r_1| > 1$, and such $h\lambda$ must be outside $\mathcal{R}_A$. By the same token, if $h\lambda$ is real, negative and small, then $|r_1| < 1$. However, it is possible that the stability polynomial possesses another root r_ν, $\nu \neq 1$, such that

$$r_\nu \to \theta_\nu, \quad |\theta_\nu| = 1, \quad \text{as} \quad h\to 0,$$

and that $|r_\nu| > 1$ when $h\lambda$ is real negative and small. This is precisely what happens in the case of Simpson's rule, (5.3), for which

$$\pi(r;\, h\lambda) = (1 - h\lambda/3)r^2 - (4/3)h\lambda r - (1 + h\lambda/3).$$

We easily find that

$$r_1 = 1 + h\lambda + O(h^2); \quad r_2 = -1 + h\lambda/3 + O(h^2),$$

so that, if Re $h\lambda$ is small and positive, $|r_1| > 1$, whereas if Re $h\lambda$ is small and negative $|r_2| > 1$. Thus Simpson's rule has no interval of absolute stability, at least in the neighbourhood of the origin. More precisely, if we apply the boundary locus method, we find that $\partial\mathcal{R}_A$ is the segment of the imaginary axis from $-\sqrt{3}j$ to $+\sqrt{3}j$. For $h\lambda$ on this segment, the roots of $\pi(r, h\lambda)$ lie on the unit circle; for all other $h\lambda$, $\pi(r;\, h\lambda)$ has a root outside the unit circle. It follows that $\mathcal{R}_A$ is empty. Recall that Simpson's rule is an optimal method (Definition 9). All optimal methods have empty regions of absolute stability; such methods cannot be recommended as general procedures.

It is clear that whenever $\rho(\theta)$ has several essential roots, then there is a danger that the region $\mathcal{R}_A$ will be empty. Such a danger does not, however, always materialize, witness the following example due to Stetter (1965):

$$y_{n+2} - y_n = \tfrac{1}{2}h(f_{n+1} + 3f_n),$$
$$\pi(r; h\lambda) = r^2 - \tfrac{1}{2}h\lambda r - (1 + \tfrac{3}{2}h\lambda), \tag{7.1}$$

whence

$$r_1 = 1 + h\lambda + O(h^2); \quad r_2 = -1 - \tfrac{1}{2}h\lambda + O(h^2). \tag{7.2}$$

For small positive Re $h\lambda$, $|r_1| > 1$, $|r_2| > 1$, while for small negative Re $h\lambda$, $|r_1| < 1$, $|r_2| < 1$, and clearly $\mathscr{R}_A$ is not empty.

Nevertheless, if we wish to be certain that the region of absolute stability will not be empty, we simply choose a method for which the principal root is the only essential root, that is, choose a method which satisfies the strong root condition (Definition 3).

Definition 11 A method of class (1.2) is said to be strongly stable if it is consistent and satisfies the strong root condition.

Following this train of thought to its conclusion, one feels assured of a substantial region of absolute stability if one chooses $\rho(\theta)$ so that all of the extraneous roots are at the origin, that is, chooses

$$\rho(\theta) = \alpha_k(\theta^k - \theta^{k-1}).$$

This defines the class of Adams methods (though usually the term is applied only to LMM). Explicit LMM of Adams type are known as Adams-Bashforth methods (AB), and implicit LMM of Adams type are known as Adams-Moulton methods (AM); their combination in PC form are known as Adams-Bashforth-Moulton methods (ABM). Several modern algorithms, such as the non-stiff version of Gear's method (Gear, 1971) are based on ABM methods. A useful study of the absolute stability properties of ABM methods has been made by Hall (1974).

8. COMPARISON OF METHODS

It is clearly impracticable to compare $\mathscr{R}_A$ for different methods on a quantitative basis; instead we list in Table 2 the intervals of absolute stability for some popular methods.

Method		Order	Interval
Euler	Exp	1	$(-2, 0)$
Back Euler	Imp	1	$(-\infty, 0) \cup (2, \infty)$
Trapezoidal	Imp	2	$(-\infty, 0)$
4-stage RK	Exp	4	$(-2.78, 0)$
2-stage RK	Imp	4	$(-\infty, 0)$
AB	Exp	4	$(-0.3, 0)$
AM	Imp	4	$(-3.0, 0)$
ABM	PEC	4	$(-0.16, 0)$
ABM	PECE	4	$(-1.25, 0)$
ABM	$P(EC)^2$	4	$(-0.9, 0)$
CKAM	PECE	4	$(-2.48, 0)$
KMAM	PEC	4	$(-0.78, 0)$
KAM	PECE	4	$(-1.8, 0)$
AM	Imp	8	$(-0.5, 0)$
ABM	PECE	8	$(-0.4, 0)$
KAM	PECE	8	$(-0.6, 0)$

Key

CKAM: Crane and Klopfenstein (1965) derived a predictor which when used with AM in PECE mode gives improved absolute stability.

KMAM: Klopfenstein and Millman (1968) derived a similar predictor for use with AM in PEC mode.

KAM: Krogh (1966) derived predictors for use with AM in PECE mode, which have good stability and require modest storage.

Table 2

All of the methods listed can be found in Lambert (1973).

Comments

1) Explicit methods (including PC) have finite intervals.

2) Implicit methods usually have larger intervals than corresponding explicit methods - but compare KAM with AM of order 8.

3) Usually (for PC methods, but not for RK) the higher the order, the smaller the interval.

4) PECE methods usually have larger intervals than corresponding PEC or $P(EC)^2$ methods.

5) 4-th order explicit RK calls for 4 evaluations per step, while PECE calls for 2. Thus, for the same effort, we can afford to use a steplength for PECE which is half that for RK. This effectively doubles the stability interval for PECE compared with RK.

6) The literature contains very little information on stability regions for hybrid methods; they would appear to be substantially smaller than for comparable PC methods.

9. RELATIVE STABILITY

We return to equation (6.8),

$$r_1 = e^{h\lambda} + O(h^{p+1}).$$

Since the solution of the test problem $y' = \lambda y$, $y(a) = y_0$ is $y(x) = y_0 e^{\lambda(x-a)}$, and $r_1^n = e^{nh\lambda} + O(h^p) = e^{\lambda(x_n-a)} + O(h^p)$, it could be claimed that we should not worry about $|r_1|$ being greater than 1 when Re $h\lambda$ is positive, for the error is then merely growing at the same rate as the solution. What would be unacceptable would be for the error to grow at a faster rate than the solution. We can therefore frame a definition of <u>relative stability</u>, which would require that $|r_\nu| < |r_1|$, $\nu = 2,3,\dots,k$. Alternatively, in view of (6.8), we could demand $|r_\nu| < |e^{h\lambda}| = 1,2,\dots,k$.

<u>Definition 12</u> The method (1.2) is said to be <u>relatively stable</u> for a given $h\lambda$ if, for that $h\lambda$, all roots of the stability polynomial are

less than $|e^{h\lambda}|$ in modulus. A region $\mathcal{R}_R$ of the complex plane is said to be a <u>region of relative stability</u> of (1.2) if (1.2) is relatively stable for all $h\in\mathcal{R}_R$. (There are many variants of this definition.)

This particular definition has the advantage that it permits us to modify the boundary locus method to find $\mathcal{R}_R$. If we write $r = Z|e^{h\lambda}|$, then a point $h\lambda$ lies on $\partial\mathcal{R}_R$ if $|Z| = 1$, that is, if there exist real ψ such that

$$\pi(|e^{h\lambda}|e^{j\psi};\ h\lambda) = 0, \quad j^2 = -1,$$

which equation now replaces (2.9).

Relative stability is not concerned solely with the case Re $h\lambda > 0$. Consider the method (7.1); it is clear from (7.2) that for small negative Re $h\lambda$, $|r_2| > |e^{h\lambda}|$, and the method is relatively unstable. Here the error is decaying, but less quickly than the solution of the test problem.

Many papers written in the 1960's, some of which have already been referred to, quote regions of relative stability. More recently, less interest has been shown in this topic, probably because of increasing interest in the problem of stiffness, to which absolute, and not relative, stability is pertinent.

3

ERROR ESTIMATION FOR INITIAL VALUE PROBLEMS

1. THE ANALYTIC PROBLEM

Consider the initial value problem

$$\left.\begin{aligned} &y'(x) - f(x,y(x)) = 0, \qquad x \in [a,b] \\ &y(a) = y_0 , \end{aligned}\right\} \tag{1.1}$$

where $\quad y(x) = \{y_1(x), \ldots, y_s(x)\}.$

We assume that f is Lipschitz continuous in y, i.e. that L exists so that

$$\|f(x,y_1) - f(x,y_2)\| \leq L\|y_1-y_2\|, \tag{1.2}$$

for all $x \in [a,b]$ and for all y_1, y_2 in the region of interest. This condition ensures that (1.1) has a unique solution (Henrici (1962).

Let us first consider the effect of perturbations on the analytic problem (1.1). Suppose that z(x) satisfies

$$\left.\begin{aligned} &z'(x) - f(x,z(x)) = \theta\, \delta(x), x \in [a,b], \\ &z(a) = y_0 + \theta\, \delta_0 , \end{aligned}\right\} \tag{1.3}$$

where θ is small.
On assuming that

$$z(x) = y(x) + \theta\, e(x) + O(\theta^2) \tag{1.4}$$

and using Taylor's theorem we get from (1.3)

$$y'(x) + \theta\, e'(x) - f(x,y(x)) - f_y(x,y(x)).\theta\, e(x) = \theta\, \delta(x) + O(\theta^2),$$

$$y(a) + \theta\, e(a) = y_0 + \theta\, \delta_0 + O(\theta^2).$$

Hence e(x) must satisfy the linear differential equation

$$\left.\begin{aligned} e'(x) - f_y(x,y(x))\, e(x) &= \delta(x), \\ e(a) &= \delta_0. \end{aligned}\right\} \tag{1.5}$$

That is if y and z are defined by (1.1) and (1.3) and e(x) by (1.5) then (1.4) holds.

We see later that errors due to solving (1.1) approximately, satisfy similar equations.

The solution of (1.5) is

$$e(x) = E(a,x)\,\delta_0 + \int_a^x E(u,x)\,\delta(u)du, \tag{1.6}$$

where

$$E(u,x) = \exp\left[\int_u^x f_y(t,y(t))dt\right]. \tag{1.7}$$

Note that if there are s equations $f_y(t,y(t))$ is an $s \times s$ matrix (the Jacobian of f) and so is E(u,x). In this case the exponential is defined by its infinite series which is always convergent.

From (1.6) we see that the effect of a perturbation $\delta(u)$ at u depends on E(u,x) which may be greater or less than one, and be an increasing or decreasing function. If the differential equation is $y' = \lambda y$ so that $f_y = \lambda$ then $E(u,x) = \exp(\lambda(x-u))$. If $\lambda > 0$ the effect of an error near u on the global error at x increases as x increases. If $\lambda < 0$ the opposite occurs. More complicated types of behaviour are possible in other equations.

This behaviour is illustrated by the characteristic curves of the differential equation. The set of characteristic curves of $y' - f(x,y) = 0$ is the set of solution curves of $y' - f(x,y) = 0$, $y(a) = y_0$ for all values of y_0. The effect of a perturbation is to push the solution trajectory from one of these curves to an adjacent one. In figure 1 we see the effect of a non-zero perturbation occurring between u and $u+\delta u$ and also of a small perturbation which is positive throughout the range.

A small local error near a causes a large global one at u_1 but one of the same order as the local error at u_2. A local error near u_1 causes a much smaller global error at u_2, measuring the size of the local error by its integral.

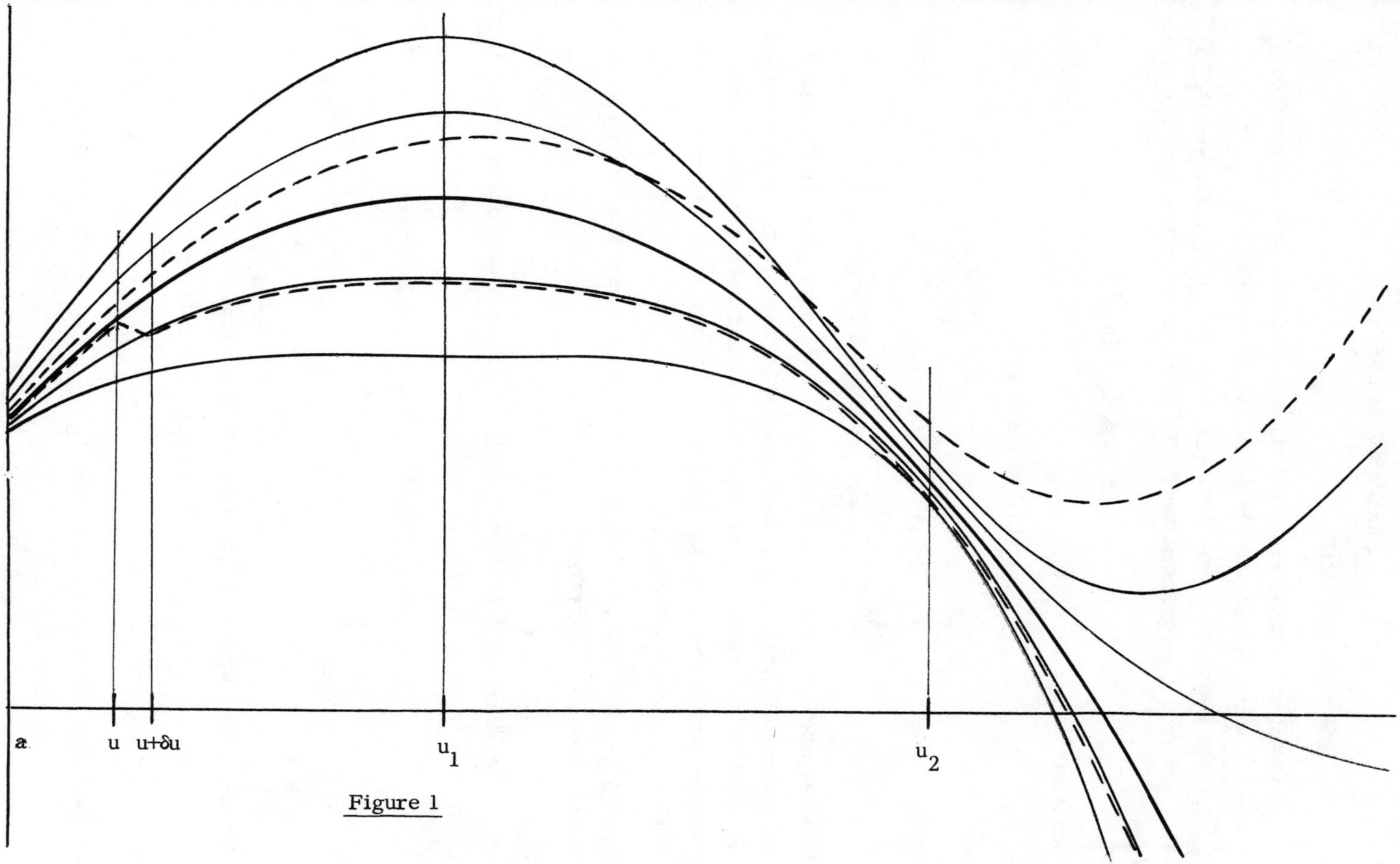

Figure 1

2. DISCRETE APPROXIMATIONS

Consider a k-step algorithm which calculates an approximation y_{n+k} to $y(x_{n+k})$ using the previously calculated approximations y_n, y_{n+1}, ..., y_{n+k-1}; the starting values, i.e. approximations to y_0, y_1, ..., y_{k-1}, being got from a separate method. We consider a uniform step $h = h_N$ so that

$$x_i = a + ih, \; h = h_N = (b-a)/N.$$

Such a method can be written

$$y_n = s_n(h), \quad 0 \leq n < k,$$

$$\sum_{i=0}^{k} \alpha_i y_{n+i}/h - \phi(x_n; y_{n+k}, \ldots, y_n; h) = 0, \qquad (2.1)$$

$$0 \leq n \leq N-k.$$

Examples, making it clear that ϕ depends on f, were given in Chapter 2. Definitions of convergence, local discretization error, consistency and zero-stability have been given previously and theorems relating these concepts have been stated in Chapter 2. We now state and prove the simplest of such theorems.

Theorem 1. If a method of form (2.1) is consistent of order p and zero-stable, then it is convergent of order p.

Proof

Equation (2.1) holds for $\{y_n\}$ and from the definition of the local truncation error $\{d_n\}$, $\{y(x_n)\}$ satisfies

$$y(x_r) - s_r(h) = d_r, \qquad 0 \leq r < k,$$

$$\sum_{i=0}^{k} \left\{\alpha_i y(x_{n+i})/h\right\} - \phi(x_n; y(x_{n+k}), \ldots, y(x_n); h) = \rho'(1)\, d_{n+k},$$

$$0 \leq n \leq N-k.$$

Now using the definition of zero-stability, with

$$'z_n^*' = y_n, \qquad 'z_n' = y(x_n), \qquad '\delta_n^*' = 0, \quad 0 \leq n \leq N,$$

$$'\delta_n' = d_n, \; 0 \leq n < k \qquad \text{and} \qquad '\delta_{n+k}' = \rho'(1)\, d_{n+k}, \quad 0 \leq n \leq N-k,$$

we find that there exist constants h_0 and S such that for all $h \in (0, h_0]$, and $0 \leq n \leq N$

$$\|y_n - y(x_n)\| \leq S\, d,$$

whenever

$$\|d_n\| \leq d, \text{ for } 0 \leq n \leq N.$$

If the method is consistent of order p, then $\max\|d_n\| = O(h^p)$, and so there exist K and h_1 such that

$$\|d_n\| \leq K h^p, \text{ for } 0 < h \leq h_1.$$

On taking $d = K h^p$ we find

$$\|y - y(x_n)\| \leq S K h^p,$$

whenever

$$0 < h \leq \min (h_0, h_1).$$

That is, the method is convergent of order p.

This theorem is of interest because it is usually easy, using Taylor expansions, to find the order of consistency of a method, or to derive conditions on the parameters occurring in it for a method to be of a particular order, and the stability is often easy to check using the 'root condition' of Chapter 2.

I now want to show that if

$$d_n = s_n(h) - y(x_n) = O(h^{p+1}), \qquad 0 \leq n < k, \tag{2.2}$$

and

$$d_{n+k} = h^p\psi(x_{n+k}) + O(h^{p+1}), \qquad 0 \leq n \leq N-k, \tag{2.3}$$

then

$$y_n = y(x_n) + h^p e(x_n) + O(h^{p+1}), \qquad 0 \leq n \leq N, \tag{2.4}$$

where

$$\begin{aligned} &e'(x) - f_y(x,y(x))\, e(x) = -\psi(x), \\ &e(a) = 0. \end{aligned} \tag{2.5}$$

Notice that the conditions imply that the method is consistent of order p but that the initial conditions are accurate to order p+1.

Substituting $y(x_n) + h^p e(x_n)$ into the left hand side of (2.1), and assuming sufficient differentiability,

$$\sum \alpha_i (y(x_{n+i}) + h^p e(x_{n+i}))/h - \phi(x_n;\, y(x_{n+k}) + h^p e(x_{n+k}), \ldots;\, h)$$

$$= \left\{ \sum \alpha_i\, y(x_n)/h - \phi(x_n;\, y(x_{n+k}), \ldots;\, h) \right\}$$

$$+ h^p \sum \alpha_i \, e(x_{n+i})/h - h^p \sum_{i=0}^{k} \frac{\partial \phi(x_n;\, y(x_{n+k}),\, \ldots;\, h)}{\partial y(x_{n+i})} e(x_{n+i}) + O(h^{2p})$$

$$= \rho'(1)\, h^p \left\{\psi(x_{n+k}) + e'(x_{n+k}) - f_y(x_{n+k},\, y(x_{n+k}))\, e(x_n)\right\} + O(h^{p+1})$$

$$= O(h^{p+1}).$$

To get the first term of the second line we have used the definition of discretization error and (2.3). For the second we have expanded $e(x_{n+i})$ about x_{n+k} and used (2.4) and the fact that $\rho'(1) = \sum i\, \alpha_i$, and for the third term, after Taylor's theorem, we have made use of an equation similar to (2.2 (iii)) of Chapter 2, but which involves derivatives with respect to y. The final line follows from (2.9).

We also have, for $0 \leq n < k$,

$$y(x_n) + h^p\, e(x_n) = s_n(h) - (s_n(h) - y(x_n)) - h^p\, e(x_n)$$
$$= s_n(h) + O(h^{p+1}).$$

Using (2.6) and noting that since $e(a) = 0$ we have

$$e(x_n) = O(h), \quad 0 \leq n \leq k.$$

Thus y_n and $y(x_n) + h^p\, e(x_n)$ satisfy sets of equations whose right hand sides differ by $O(h^{p+1})$ and we can prove that if the method is zero stable y_n and $y(x_n) + h^p\, e(x_n)$ differ by $O(h^{p+1})$. The proof is similar to that of Theorem 1.

This result means that a method for which the principal part of the local discretization error is $h^p\, \psi(x_n)$ at x_n has global discretization error with principle part $h^p\, e(x_n)$ and the deviation from the true solution is the same as that due to a perturbation $h^p\, \psi(x_n)$ in the analytic equation, provided that the initial conditions are correct to order p+1. (Compare (1.3) and (1.4) with (2.5) and (2.4).)

The analysis can be extended to give useful results for some variable-step and variable order methods; for a detailed treatment, see Stetter (1973).

We also see that if by some means we control the local discretization error to be less than, say 10^{-5} the error in the solution will be less than that caused by perturbations of order 10^{-5} in $f(x,y)$. That is, our result can be expressed in backward error analysis form. This approach will often provide the best way to explain an algorithm's error control strategy to a user.

A local discretization error at one point affects the error in the solution at subsequent points in just the same way as does a perturbation in the analytic problem. The ways such errors increase or decrease with increasing x was discussed in section 1.

In section 1 we also wrote down the solution of the linear differential equation for $e(x)$, which is the quantity that we are really interested in. Knowledge of $e(x)$ depends on knowledge of the s^2 components of $E(u,x)$ and the amount of work involved makes these impractical to obtain, unless the Jacobian is already available, as it will be in some methods for integrating stiff equations. We have usually to be content with the local error, which can be estimated without too much difficulty. There is however one particular class of cases investigated by Stetter ((1973) pp.161 and 297) in which the differential equation for $e(x)$ can be easily solved. I will merely give an example of this.

Suppose that $\psi(x) = C(y^{(p+1)}(x) - f_y(x,y(x)).y^p(x))$ and that the initial condition is

$$e(a) = -C\, y^p(a).$$

Then $e(x) = Cy^p(x)$, as is easily verified; so that an approximation to the global error is immediately available by numerical differentiation from the approximate solution.

For this solution to be applicable it is necessary that the starting values of the approximate solution satisfy

$$s_n(h) = y(x_n) - C\, h^p\, y^{(p)}(x_n) + O(h^{(p+1)}).$$

Stetter has shown that to every corrector scheme there corresponds a predictor of degree one less so that the error of the method has this special form, and that suitable Runge-Kutta schemes can also be found. The existence of such schemes was implied by a paper of Butcher (1969).

When (2.4) and (2.5) hold, they can be used to link the local error l_n at x_n (Chapter 1, definition 4.3) with the local discretization error. The relationship is

$$l_n = h(d_n + O(h^{p+1})), \qquad k \leq n \leq N.$$

Also on the assumption that (2.4) and (2.5) hold, one can estimate the global error at a fixed point c from the approximate solutions

$y_M(h)$ and $y_{2M}(h/2)$ at c, calculated at steplengths h and h/2 respectively, where $Mh = c-a$. We have

$$y_M(h) = y(c) + h^p e(c) + O(h^{p+1})$$

and

$$y_{2M}(h/2) = y(c) + (h/2)^p e(c) + O(h^{p+1}).$$

Hence

$$y_{2M}(h/2) - y(c) = [y_M(h) - y_{2M}(h/2)]/[2^p-1] + O(h^{p+1}).$$

In other words the error at step h/2 is about 2^{-p} times the difference between solutions at steps h and h/2.

This gives us the most commonly used elementary way to choose a suitable step length for a non-automatic routine; calculate the solution at a sequence of smaller and smaller steps until two solutions agree to the required accuracy.

If an automatic routine is being used one would expect the global error to decrease linearly with the supplied accuracy parameter, but this may not be so, since it will depend on exactly how the choice of step length and order is done.

In a one-step method which changes step length at each step to an estimated optimum, one would expect it to be so, but in a similar routine which adjusted the step length by halving and doubling only, the graph of global error against accuracy parameter could be a piecewise constant function in which the error changed by a factor 2^{-p} at each discontinuity.

Care is therefore needed in trying to estimate the global error of automatic routines.

3. THE FORM OF THE LOCAL DISCRETISATION ERROR

Consider an m-stage k-step method in which when calculating $y_n \doteq y(x_n)$, (m-1) approximations to the values $y(x_n + \theta_i h)$ are also found. (Thus for fourth order Runge-Kutta, m = 4. For a simple predictor-corrector method m = 2.)

Formulae used are of the form

$$y_{q,n+k} = \sum_{i=0}^{k-1} (-\alpha_{q,i} y_{n+i} + h\,\beta_{q,i}\, f(y_{n+i})) + h \sum_{r=1}^{q-1} a_{q,r} f(y_{r,n+k}), \quad (3.1)$$

$$0 \le n \le N-k, \qquad 1 \le q \le m,$$

$$y_{n+k} = y_{m,n+k}, \tag{3.2}$$

with starting values given by

$$y_n = s_n(h), \qquad 0 \le n < k. \tag{3.3}$$

(For convenience we have suppressed the argument x in f. This should not cause any confusion.)

The form (3.1) can be seen to apply to Runge-Kutta, predictor-corrector and to most hybrid methods provided that the argument used for f in the $\beta_{qi}f$ term is y_{n+1} and not y_{n+i,m_1} for some $m_1 \neq m$.

Equations (3.1) are used to find $y_{1,n+k}, y_{2,n+k}, \ldots, y_{m,n+k}$ in turn but only $y_{m,n+k} (= y_{n+k})$ and $f(y_{n+k})$ are carried forward to subsequent steps, the other intermediate quantities being forgotten! So y_{n+k} is a function only of $y_n, y_{n+1}, \ldots, y_{n+k-1}$. The intermediate quantities can be easily eliminated and the method written in form (2.1).

In order to find the local discretization error d_n we need the error in this final form with the intermediate quantities eliminated, but it is not convenient to do the analysis in that way.

We therefore define $t_{q,n+k}$ by

$$y(x_{n+k} + c_q h) - t_{q,n+k} =$$

$$\sum_{i=0}^{k-1} (-\alpha_{q,i} y(x_{n+i}) + h\,\beta_{q,i} f(y(x_{n+i}))) + \sum_{r=1}^{q-1} h\, a_{q,r}\, f(y(x_{n+k}+c_q h)), \tag{3.4}$$

$$1 \le r \le m,$$

where $c_1, c_2, \ldots, c_m$ can be arbitrary, provided that $c_m = 0$, but are usually chosen to make $t_{1,n+k}, t_{2,n+k}, \ldots$ of order at least h^2.

Using $f(y(x)) = y'(x)$ and expanding (3.4) by Taylor series about x_n or any other convenient point, we easily find an expansion

$$t_{q,n+k} = \sum_{u=p_q}^{p} d_{q,k}\, h^u y^{(u)}(x_n) + O(h^{p+1}), \tag{3.5}$$

where the $d_{q,k}$ are numerical coefficients, and $p_1, p_2, \ldots$ are the orders of the first, second, ... stages.

Thus the $t_{q,n+k}$, the errors in $y_{q,n+k}$ assuming no propagation of error from stage to stage, are easily found.

We now go on to find $T_{q,n+k}$ the errors in $y_{q,n+k}$ allowing for propagation of error in the present step only. Propagation of error from step to step is dealt with by stability theory as described briefly in section 2.

Define $T_{q,n+k}$ by

$$y(x_{n+k}+c_q h) - T_{q,n+k} = \sum_{i=0}^{k-1} (-\alpha_{q,i} y(x_{n+i}) + \beta_{q,i} hf(y(x_{n+i}))) + h \sum_{r=1}^{q-1} a_{q,r} f(y(x_{n+k}+c_r h) + T_{r,n+k}), \quad 1 \le q \le m. \tag{3.6}$$

$$\left.\begin{aligned} \text{So } T_{1,n+k} &= t_{1,n+k}, \\ T_{2,n+k} &= t_{2,n+k} + ha_{21} f_y(y(x_{n+k}+c_q h))t_{1,n+k} \\ &\quad + \tfrac{1}{2}h\, a_{21} f_{yy}(\ \ )(t_{1,n+k})^2 + \ldots, \\ T_{3,n+k} &= t_{3,n+k} + ha_{32} f_y(\ \ )\{t_{2,n+k} + ha_{21} f_y(\ \ )t_{1,n+k} + \ldots\} \\ &\quad + ha_{31} f_y(\ \ )t_{1,n+k} - \ldots . \end{aligned}\right\} \tag{3.7}$$

Hence $T_{n+k} = T_{m,n+k}$ can be obtained for such explicit methods in a straightforward manner although its form may be complicated.

Since our method (3.1), (3.2) is a particular case of (2.1) we can write it in the form

$$(y_{n+k} + \sum_{i=0}^{k-1} \alpha_{m,i} y_{n+1})/h - \phi(x_n;\ y_{n+k-1},\ldots,y_n;\ h) = 0 \tag{3.8}$$

and in this notation

$$y(x_{n+k}) - T_{n+k} + \sum_{i=0}^{k-1} \alpha_{m,i} y(x_{n+i}) - h.\ \phi(x_n;\ y(x_{n+k-1}),\ldots,y(x_n);\ h) = 0. \tag{3.9}$$

Hence

$$d_{n+k} = T_{n+k}/h\rho'(1), \tag{3.10}$$

where

$$\rho(\theta) = \theta^k + \sum_{i=0}^{k-1} \alpha_{m,i}\, \theta^i \tag{3.11}$$

and if the method is of order p we get

$$d_{n+k} = T_{n+k}/h\ \rho'(1) = h^p \psi(x_{n+k}) + O(h^{p+1}). \tag{3.12}$$

<u>Example</u>: Second order Runge Kutta.

$k = 1$, $m = 2$, $p = 2$.

Since this is an explicit one step method $\rho(\theta) = \theta - 1$ and $\rho'(1) = 1$, equations (3.1) and (3.2) reduce to

$$y_{1,n+1} = y_n + hf(y_n),$$
$$y_{2,n+1} = y_n + h\,\tfrac{1}{2}f(y_n) + h\,\tfrac{1}{2}f(y_{1,n+1}),$$
$$y_{n+1} = y_{2,n+1}$$

and $s_0(h) = y_0$.

We take $c_1 = c_2 = 0$,

$$t_{1,n+1} = y(x_{n+1}) - y(x_n) - h\, f(y(x_n))$$
$$= \left[y + hy' + \frac{h^2}{2} y'' + \ldots + -y - hy'\right]_{x_n},$$

i.e. $t_{1,n+1} = \dfrac{h^2}{2} y''(x_n) + O(h^3)$,

$$t_{2,n+1} = y(x_{n+1}) - y(x_n) - \tfrac{1}{2}hf(y(x_n)) - \tfrac{1}{2}hf(y(x_{n+1}))$$
$$= \left[y + hy' + \frac{h^2}{2} y'' + \frac{h^3}{6} y''' - y - \tfrac{1}{2}hy' - \tfrac{1}{2}h\left(y' + hy'' + \frac{h^2}{2}y'''\right) + O(h^4)\right]_{x_n},$$

i.e. $t_{2,n+1} = -\dfrac{1}{12} h^3\, y'''(x_n) + O(h^4)$.

So $T_{2,n+1} = -\dfrac{1}{12} h^3\, y'''(x_n) + \tfrac{1}{2}h.\ f_y(y(x_n))\dfrac{h^2}{2}\, y''(x_n) + O(h^3)$

and $d_{n+1} = T_{2,n+1}/h = -\dfrac{1}{12} h^2(y'''(x_n) - 3f_y(y(x_n))y''(x_n)) + O(h^3)$.

Hence the order of the method is 2 and

$$y_n = y(x_n) + h^2\, e(x_n) + O(h^3),$$

where $e'(x) - f_y(y(x))\, e(x) = \dfrac{1}{12}\,(y'''(x) - 3f_y(y(x))y''(x))$,

$$e(a) = 0.$$

4. ESTIMATION OF LOCAL DISCRETIZATION ERROR OF ONE-STEP METHODS

There are three main methods that can be used:

i) Extrapolation. The integration from x_n to x_{n+1} is done twice; as one step of length h and as two steps of length h/2. Then $d_n = 1/(h(1-2^{-p})) \times$ (difference between the two y-values).

ii) Embedding Methods. Each step is integrated twice, by a pth order and a $(p+1)^{th}$ order method. The difference between the values obtained gives an estimate of d_n.

iii) Multistep Estimates. A multistep formulae which is of order (p+1) is evaluated to provide the required estimate

$$d_n = \left\{ \sum_{i=-k+1}^{1} \alpha_i y_{n+i}/h - \beta_i f(y_{n+i}) \right\} / \rho'(1).$$

iv) Hybrid Methods which are mixtures of ii) and iii) are also available.

Extrapolation has been the most popular method, but it is the least efficient. It can be used in two ways; after the integration has been done as one-step and as two-steps either the one-step or the two-step estimate of $y(x_{n+1})$ can be used as the value on which the subsequent computation is based. The following analysis applies if the one-step value is used, but it is perhaps preferable to use the two-step value for which a similar treatment holds.

In the one-step explicit case (2.1) reduces to

$$y_0 = y_0,$$

$$(y_{n+1} - y_n)/h - \phi(x_n, y_n, h) = 0, \qquad 0 \le n < N, \tag{4.1}$$

and $\rho(\theta) = \theta - 1$, $\rho'(1) = 1$, $\alpha_0 = -1$.

If this is consistent of order p, since one-step methods are always stable

$$y_n = y(x_n) + \varepsilon_n, \tag{4.2}$$

where

$$\varepsilon_n = O(h^p).$$

Write $\bar{y}_{n+1}$ for the result of taking two-steps of length h/2 from y_n. Then defining

$$\bar{\phi}(x_n,y_n,h) = \tfrac{1}{2}\{\phi(x_n,y_n,\tfrac{h}{2}) + \phi[x_n + \tfrac{1}{2}h, y_n + \tfrac{1}{2}\phi(x_n,y_n,\tfrac{h}{2}), \tfrac{h}{2}]\}, \qquad (4.3)$$

we can write

$$\bar{y}_{n+1} = y_n + h\,\bar{\phi}(x_n, y_n, h).$$

A simple analysis like that of section 3 now shows that if $\bar{T}_{n+1}$ is the truncation error for $\bar{\phi}$, in an obvious notation

$$\bar{T}_{n+1} = (\tfrac{1}{2})^{p+1}\, T_{n+\frac{1}{2}} + \{1 + \tfrac{h}{2}\,\phi_y(\quad)\}\,(\tfrac{1}{2})^{p+1}\, T_{n+1} + O(h^{p+2})$$

and using

$$T_{n+1} = h^{p+1}\,\psi(x_{n+1}) + O(h^{p+2}),$$

we get

$$\bar{T}_{n+1} = (\tfrac{1}{2})^{p}\, h^{p+1}\,\psi(x_{n+1}) + O(h^{p+2}), \qquad (4.4)$$

assuming that $\psi(x)$ is sufficiently smooth.

Now we make use of (3.9), (4.2) and a generalisation of (2.2 iii) of Chapter 2.

$$\begin{aligned} y_{n+1} &= y_n + h\,\phi(x_n,y_n, h) \\ &= y(x_n) + \varepsilon_n + h\,\phi(x_n,y(x_n)+\varepsilon_n, h) \\ &= y(x_n) + h\,\phi(x_n,y(x_n), h) + h\,\phi_y(\quad)\,\varepsilon_n + \varepsilon_n + O(h^{p+2}). \end{aligned}$$

i.e. $$y_{n+1} = y(x_{n+1}) - T_{n+1} + f_y(\quad)\,h\,\varepsilon_n + \varepsilon_n + O(h^{p+2}). \qquad (4.5)$$

Similarly

$$\bar{y}_{n+1} = y(x_{n+1}) - \bar{T}_{n+1} + f_y(\quad)\,h\,\varepsilon_n + \varepsilon_n + O(h^{p+2}).$$

Subtracting and using the approximations (3.12) and (4.4)

$$\bar{y}_{n+1} - y_{n+1} = h^{p+1}\,\psi(x_{n+1})\,(1-2^{-p}) + O(h^{p+2}),$$

and the local discretization error of the method using step h is

$$\begin{aligned} d_{n+1} &= h^{p}\,\psi(x_{n+1}) + O(h^{p+1}) \\ &= \frac{(\bar{y}_{n+1} - y_{n+1})}{h(1-2^{-p})} + O(h^{p+1}). \end{aligned} \qquad (4.6)$$

One can also easily show that the discretization error of the method using step h/2, that is continuing the calculation from $\bar{y}_{n+1}$ instead of from y_{n+1} is just (4.6) reduced by a factor 2^{-p}.

<u>Embedding Methods</u> are a recent innovation due to England, Fehlberg and Shintani.

The basic idea is to perform the step from x_n to x_{n+1} with a pth order and a $(p+1)^{th}$ order method, to get an estimate of the error in the pth order integration. We thus have, if $\bar{\phi}$ determines the $(p+1)^{th}$ order method

$$y_{n+1} = y_n + h\,\phi(x_n, y_n, h)$$

and

$$\bar{y}_{n+1} = y_n + h\,\bar{\phi}(x_n, y_n, h).$$

So y_{n+1} satisfies (4.5) and $\bar{y}_{n+1}$ satisfies similarly

$$\bar{y}_{n+1} = y(x_{n+1}) - \bar{T}_{n+1} + f_y(\quad).\,h\,\varepsilon_n + \varepsilon_n + O(h^{p+2}).$$

Subtracting

$$\bar{y}_{n+1} - y_{n+1} = T_{n+1} - \bar{T}_{n+1} + O(h^{p+2})$$

$$= T_{n+1} + O(h^{p+2}).$$

So $$d_{n+1} = (\bar{y}_{n+1} - y_{n+1})/h. \qquad (4.7)$$

Again one has the choice of continuing the calculation from either the pth or $(p+1)^{st}$ order value.

A fourth order method needs 4 function evaluations and a fifth order one 6 evaluations. If the methods are unrelated only the first evaluation is common so 9 evaluations are needed per step for a fourth order method with error estimate.

Embedding methods are ones where a pth order method is obtained as a by-product of a $(p+1)^{st}$ order one, in this way both a fourth and fifth order estimate can be obtained from a method using six function evaluations - a saving of 3 evaluations.

The algebra involved in obtaining such methods is formidable, but they have been obtained by England (1967, 1969), Shintani (1965, 1966a, 1966b) and Fehlberg (1968, 1969a, 1969b). Fehlberg has produced a set of such methods, some details of which are given below.

p	1	2	3	4	5	6	7	8	9
Least no. of stages for order p	1	2	3	4	6	7	9	≥10	≥11
No. of stages used by Fehlberg for order p and (p+1) method	2	3	5	6	8	10	13	17	-
No. of stages for independent methods of order p and p+1	2	4	6	9	12	15	≥18	≥20	

One might expect that for $p = 3$ such a method would be possible with 4 stages, but England has proved that this is impossible.

A sophistication possible in such methods is that since more than the minimum number of function evaluations, (or stages) are available for the pth order method, it may be possible for that method to have a much smaller error bound than usual.

The well known method of Merson (1957), (see also Lambert (1973), p.131) which is used in the N.A.G. Library is a fourth order 5 stage method, which gives an error estimate valid only if f is of the form $f(x,y) = a+bx+cy$. In other cases it includes some lower order terms and so it usually over-estimates the error (Scraton (1964)). It is possible however for the error to be under-estimated (England 1969)).

The paper by Scraton also gives some nonlinear methods for estimating the error which are valid only for a single real equation (not for systems). Their use is therefore very limited.

Example: Fehlberg (1969b)

$$y_{1,n+1} = y_n + \frac{1}{4} h f(y_n),$$

$$y_{2,n+1} = y_n - \frac{189}{800} h f(y_n) + \frac{729}{800} h f(y_{1,n+1}),$$

$$y_{n+1} = y_{3,n+1} = y_n + \frac{241}{891} h f(y_n) + \frac{1}{33} h f(y_{1,n+1}) + \frac{650}{891} h f(y_{2,n+1}),$$

$$\bar{y}_{n+1} = y_{4,n+1} = y_n + \frac{233}{2106} h f(y_n) + \frac{800}{1053} h f(y_{2,n+1}) - \frac{1}{78} h f(y_{3,n+1}).$$

<u>Multistep</u> methods for error estimation are similar in principle to the previous ones. A multistep formula of order $(p+1)$ is evaluated to give the estimate of the local error.

If α_i^*, β_i^* $0 \leq i \leq k$ are the coefficients of a k-step method of order $(p+1)$ with polynomial ρ^* then we have

$$\sum_{i=0}^{k} (\alpha_i^* z(x_{n+i})/h - \beta_i^* z'(x_{n+i})) = O(h^{p+1}), \tag{4.8}$$

if z is any sufficiently differentiable function. Also

$$\rho^*(1) = \sum \alpha_i^* = 0 \text{ and } \rho^{*\prime}(1) = \sum \beta_i^*, \tag{4.9}$$

since it is consistent of order greater than one.

We must now assume that the solution of our one-step method of order p satisfies

$$y_n = y(x_n) + h^p e(x_n) + h^{p+1} e_1(x_n) + O(h^{p+1}). \qquad (4.10)$$

This is an extension of the result proved in section 2. It follows if $f(x,y)$ is smooth enough. Using (2.5) we get

$$\sum_{i=0}^{k} \left\{ \alpha_i^* y_{n+1}/h - \beta_i^* f(y_{n+1}) \right\} = \sum \left\{ \alpha_i^* y(x_{n+i})/h - \beta_i^* f(y(x_{n+i})) \right\}$$

$$+ h^p \sum \left\{ \alpha_i^* e(x_{n+i})/h - \beta_i^* f_y(y(x_{n+i}))e(x_{n+i}) \right\}$$

$$+ h^{p+1} \sum \alpha_i^* e_1(x_{n+i})/h + O(h^{p+1})$$

$$= \sum \left\{ \alpha_i^* y(x_{n+i})/h - \beta_i^* y'(x_{n+i}) \right\}$$

$$+ h^p \sum \left\{ \alpha_i^* e(x_{n+i})/h - \beta_i^* e'(x_{n+i}) \right\}$$

$$+ \sum \beta_i^* \psi(x_{n+i}) + O(h^{p+1}).$$

We have used the continuous differentiability of e_1 and $\sum \bar{\alpha}_i = 0$ to replace the e_1 term by $O(h^{p+1})$. Now using (4.8) twice with $z(x)$ replaced by $y(x)$ and $e(x)$ and noting (4.9) we get finally if $\psi(x)$ is continuously differentiable

$$\sum_{i=0}^{k} \left\{ \alpha_i^* y_{n+i} - \beta_i^* f(y_{n+1}) \right\} / \rho'(1)$$

$$= h^p \psi(x_{n+k}) + O(h^{p+1}) = d_{n+k} + O(h^{p+1}). \qquad (4.11)$$

This estimate of d_{n+k} is calculated using the values of y_{n+i} and of $f(y_{n+i})$ which are available from k previous steps. It is thus cheap to evaluate and no extra evaluations of $f(y)$ are needed. The disadvantages stem from the fact that it is a multistep estimate. We thus get all the attendant disadvantages of multistep methods. It does not apply over the first few steps of an integration or after a step change unless special formulae are calculated after each change of step. It is also inevitable that the result given by such a method should be to some extent an average result for the range $[x_n, x_{n+k}]$ and this makes it unreliable when the local error is changing rapidly.

Shampine and Watts (1971) recommend a formula of Ceschino and Kuntzman (1963) as being good of this type for estimating the errors in fourth order methods. It is

$$d_{n+1} = (11\,y_{n+1} + 27\,y_n - 27\,y_{n-1} - 11\,y_{n-2})/60\,h$$
$$- (f(y_{n+1}) + 9\,f(y_n) + 9\,f(y_{n-1}) + f(y_{n-2}))/20.$$

<u>Hybrid methods</u> combining some of the features of the embedded and multi-step methods have been considered by England (1967, 1969). He considers error estimates for fourth order Runge Kutta methods using values obtained over two Runge Kutta steps, but makes use of internal values at such steps as well as the external values y_n and $f(y_n)$. He designs the Runge-Kutta step and its error estimator simultaneously and is able to produce reliable estimates cheaply.

One of these, England (1969), equation 8, estimates the error in the step [x, x+h] after the four evaluations for [x, x+h] plus three for [x+h, x+2h] plus one additional evaluation have been made. If this error is too big, seven evaluations are wasted (that of f(y(x)) can be used again). If the estimate is all right five evaluations per step have been used.

Another, detailed in his equation 9, calculates the error over <u>two</u> steps [x,x+h], and [x+h,x+2h] at the same point of the calculation. The extra evaluation is needed only once every two steps, so the average number of evaluations per step is four and a half and seven evaluations are wasted if the estimated error is too big.

These figures compare with six evaluations per step with five wasted if the error estimate is too big for an embedding method.

In our notation, England's equation (9) is

$$y_{1,n+1} = y_0 + \tfrac{1}{2}h\ f(y_n),$$
$$y_{2,n+1} = y_0 + \tfrac{1}{4}h\ f(y_n) + \tfrac{1}{4}h\ f(y_{1,n+1}),$$

$$y_{3,n+1} = y_0 - h\, f(y_{1,n+1}) + 2h\, f(y_{2,n+1}),$$

$$y_{4,n+1} = y_0 + \tfrac{1}{6}h \left\{ f(y_n) + 4f(y_{2,n+1}) + f(y_{3,n+1}) \right\},$$

$$y_{n+1} = y_{4,n+1},$$

$$\tfrac{1}{2}(d_{n+1} + d_n) = \left\{ -f(y_0) + 4f(y_{2,n+1}) + 17f(y_{3,n+1}) - 23f(y_{n+1}) + 4f(y_{2,n+2}) - f(y_{5,n+2}) \right\}/180,$$

where

$$y_{5,n+2} = y_0 + \tfrac{1}{6}h \left\{ -f(y_0) - 96f(y_{1,n+1}) + 92f(y_{2,n+1}) - 121f(y_{3,n+1}) + 144f(y_{n+1}) + 6f(y_{1,n+2}) - 12f(y_{2,n+2}) \right\},$$

and $d_{n+1} + d_n$ and $y_{5,n+2}$ are calculated once every two steps.

5. COMPARISON OF ERROR ESTIMATES FOR ONE-STEP METHODS

We can base a comparison on several grounds i.e. a) accuracy, b) reliability, c) direct cost and d) indirect cost.

Shampine and Watts (1971) have compared extrapolation, England's method (4.1) and two multistep methods. They conclude as a result of numerical experiments that estimates based on extrapolation and on England's method of estimating over two steps (4.1) could hardly be separated in any of the tests. One multistep method was the most accurate and one the least accurate of the four tested methods, when h was very small. The multistep estimates appear to be rather less reliable, but the statistics quoted were not ideal for such uses, being merely averages. Some extreme values of the ratio of actual to estimated errors would have been useful.

In their major paper on comparison of numerical methods, Hull, Enright, Fellen, and Segwick (1972) used extrapolation only, and compared fourth, sixth, and eighth order methods. They found that with most problems the maximum rate of actual to estimated error rarely exceeded 5 but of the 75 computer runs (3 accuracies on 25 problems) a maximum ratio of 20 was exceeded on 16 problems, and a maximum ratio of 100 on 10 problems. The two highest maxima were 1120 and 1788. We are not informed whether these very large ratios occurred because the estimated error was particularly small, near a point where the actual error was changing sign.

I know of no other published information on the reliability of estimators and think that there is scope for much more work to be done on this subject.

Theoretical results indicate that extrapolation should be by far the most reliable estimator, since higher order terms in the error are estimated with very little error. (For a method of order four this error is less than 15%.)

I think I should comment here on those embedded methods which use the extra function evaluations available to produce a pth order approximation to y_n with a particularly small principal error. I expect that the reliability of the associated estimate of local discretization error to be poor, for it estimates only the terms of order p, which have been made artificially small. For moderate h the order (p+1) terms could then well be dominant.

Multistep methods are cheapest in direct costs, but force the storage of (k-1) previous results and complicate step changing, so their indirect costs are very high.

Of the other methods those of England for p=4 are cheapest and should be used for that case, otherwise embedding methods should be used, unless it is found that the extra reliability of extrapolation methods outweighs their extra cost in a particular application.

6. ESTIMATION OF LOCAL DISCRETIZATION ERROR FOR PREDICTOR CORRECTOR METHODS

A method known as Milne's device is almost universally used in this case since it is simple and efficient. It applies to the case where both predictor and corrector formulae are of order p.

In this case the error $T_{1,n+k}$ (c.f. section 3) in the first (predictor) stage, and T_{n+k} in a whole step of the method can be written in the forms,

$$T_{1,n+k} = C^* h^{p+1} y^{(p+1)}(x_{n+k}) + O(h^{p+2}),$$

$$T_{n+k} = C\, h^{p+1} y^{(p+1)}(x_{n+k}) + O(h^{p+2}).$$

These equations serve to define the constants C^* and C which in fact depend only on the coefficients of the predictor and corrector equations, respectively.

Remembering that T_{n+k} and $\bar{T}_{n+k}$ are the errors in the predicted value $\bar{y}_{n+k}$ and corrected value y_{n+k} of $y(x_{n+k})$ provided all previous values of y_r are exact, it seems that we should be able to estimate $h^{p+1}y^{(p+1)}(x)$ and hence T_{n+k} and $h^p\psi(x_{n+k})$ (c.f. 3.12) from their difference. If the previous values are subject to error the situation is not so clear, but it has been treated by Henrici (1962) on page 255 and Stetter (1973) on pages 260 and 294. The correct normalisation of all qualities occurring in the estimator is very important. Shampine (1973) questions the validity of the estimates in the general case due to his incorrect normalisation.

We assume that the errors in the starting values are of order h^{p+1} (one higher order than needed for a pth order method), that the step length values are either constant, or form a 'coherent' sequence (see below) and that the characteristic polynomials ρ^* and ρ of the predictor and corrector satisfy $\rho^*(\theta) = 0$ if $|\theta| = 1$ and $\rho(\theta) = 0$. (i.e. $\rho^*(\theta)$ must contain all the 'essential' zeros of $\rho(\theta)$). (Since the method must be stable $\rho(\theta)$ can have no zero of modulus greater than one, but this restriction does not apply to $\rho^*(\theta)$). We now get the estimate

$$d_{n+k} = h^p\psi(x_{n+k}) + O(h^{p+1}) = C\,h^p y^{(p+1)}(x_{n+k}) + O(h^{p+1})$$

$$= K(y_{n+k} - \bar{y}_{n+k})/h, \tag{6.1}$$

where

$$K = \frac{a_k^*\, C/\rho'(1)}{C/\rho'(1) - C^*/\rho^{*\prime}(1)}\,. \tag{6.2}$$

A coherent grid sequence is one derived from a piecewise constant function $\theta(x)$ on $[a,b]$ by $h_n = h\,\theta(x_n)$. The steplength thus changes at a finite number of points. For a method to be of overall order p - we must use multistep methods of order at least (p-1) at these points where the step length changes.

These formulae are so easy and cheap to use that no other method need be sought. Gear's variable order / variable step method, however, estimates $y^{(p+1)}$ in a more direct manner by differencing approximations to lower derivatives.

Hamming (1959) has suggested that in the above situation one can modify the final approximation by forming the combination

$$\left\{C\, y_{n+k}/\rho'(1) - C^*\bar{y}_{n+k}/\rho^{*\prime}(1)\right\}(C/\rho'(1) - C/\rho^{*\prime}(1)),$$

so eliminating the principal part of the error and getting a method of order $(p+1)$.

This is not to be recommended, as such a combination is equivalent to using a different corrector equation, which is of order $(p+1)$, the stability of which should be investigated. (Shampine (1973), Stetter (1973) p.296).

7. LOCAL DISCRETIZATION ERROR OF HYBRID METHODS

Little work has been done on error estimates for such methods. The principal term of the error can be much more complicated than that of predictor corrector methods, as complicated as that of one step methods.

The easiest way to estimate the local error is therefore to use the multistep method where in this case, since it will usually not be possible to match all the terms in the principal error, our multistep estimating formula must be of order $(p+1)$ the same order as for one-step methods, and one higher order than for the estimate for multistep predictor corrector which uses the predictor as the estimator. The results of section 6 now apply with $C^* = 0$. For some hybrid methods which use a predictor and corrector of the same order, the theory of section 6 applies directly.

In assessing a hybrid method it is important to consider all terms in the error. If some of the terms in the error can be written as

$$C\, h^p y^{(p+1)}(x_n) + D\, h^{p+1}\, f_y(y(x_n)).\, y^{(p+1)}(x_n), \tag{7.1}$$

it is tempting to ignore the second term. However, for $y' = \lambda y$ this term is about $Dh\lambda/C$ times the first term. We will wish to use the method for $h\lambda$ at least 0.1 and probably much larger if a high order method is being used. If the second term is to be negligible we thus need $|D\, h\lambda| < |C|$, but very often since hybrid correctors have small error constants, $|C| > |D|$ making the higher order terms very important for values of h we wish to use. For example Gragg and Stetter (1964) derive a 6th order hybrid method with $m = 3$ for which the error is

$$-0.0000183\, h^6\, y^{(7)}(x_n) + 0.00656\, h^7\, f_y(x_n) y^{(7)}(x_n) + O(h^8).$$

The higher order term in the error is dominant if $|h\lambda| > 0.00279$.

These remarks apply with even more force if the second term in (7.1) is of order p. The methods of Kohfeld and Thompson (1967) have been particularly designed to overcome this difficulty.

For more details of hybrid methods the reader should consult Lambert (1973) Chapter 5. This standard text, which concentrates on descriptions of methods and implications of the theory on them, while omitting proofs of the more complicated results, treats much of the material in this chapter in more detail. For proofs see Henrici (1962) or the recent modern comprehensive, but more difficult treatise of Stetter (1973).

4

GENERAL DISCUSSION OF IMPLEMENTATION PROBLEMS

1. INTRODUCTION

The main topic discussed in this chapter is the organisation of routines for solving the initial value problem. We are concerned with the interface between the user and the integration routine and also with the internal organisation of the routine. The traditional simple integrator is discussed as well as more sophisticated versions that allow for more control by the user over the action of the routines.

The chapter must necessarily be based to a large extent on the work of Hull and his school, and of Krogh, (see the references at the end of the book). There is also a recently published book by Shampine and Gordon (1975) which describes a variable step/variable order Adams method in theoretical and practical detail, and gives FORTRAN listings of all the programs described. The book by Gear(1971) also contains much practical information and FORTRAN listings.

2. STRUCTURE OF A SIMPLE INTEGRATOR

We first discuss in a general way the structure of a routine with a very simple interface. Internally it will probably use many steps of some variable order / variable step method to perform the integration from the given initial point to a given value of the independent variable.

In use it will probably be called repeatedly. At each call the given initial values - approximations to the solution at $a+(r-1)H$ will be replaced by approximations to their values at $a+rH$. The solution is then available for output, (or other use) at the interval H.

The action of the integration routine is easiest to understand, and hence has a better chance of behaving correctly in all circumstances, if structured programming techniques are used. Hull (1974) has looked at this particular problem in detail and suggests the following organisation

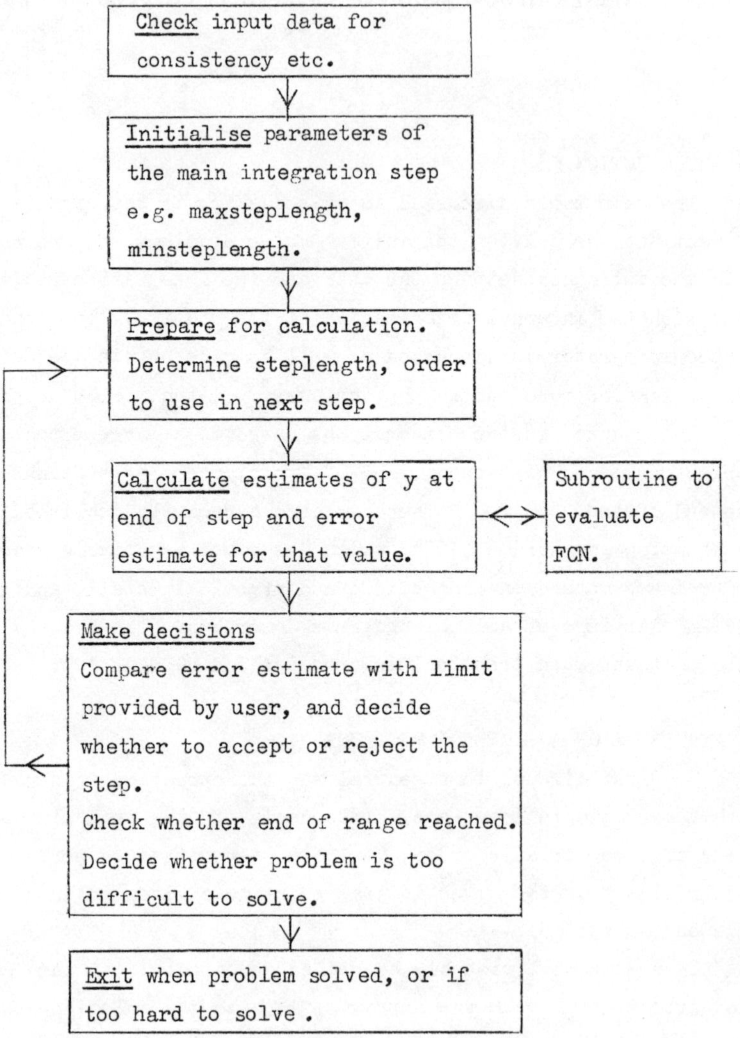

FIGURE 1

The 'calculate' block will need values of the right-hand side of the equations evaluated at a given point. This calculation is done by a subroutine provided by the user and having parameters X,Y to provide input values and an array DY to hold the results.

It should particularly be noticed that all the tests on the accuracy of a step are gathered together into the one decision block. Markers will be set by this block to inform the 'prepare' block whether the previous step is to be accepted or repeated at a smaller step-length. The accuracy tests compare either the estimated discretization error, or the estimated truncation error (which differ by a factor of h) with values provided by the user. There is much controversy as to whether it is best to use the discretization or truncation error criterion - usually described as 'error per unit step' or 'error per step' criteria. It is a gross simplification to say that the first has theoretical and the second practical advantages.

3. PARAMETER LIST OF A SIMPLE INTEGRATOR

Suppose that the analytic problem to be solved is:

$$\left.\begin{aligned} \underset{\sim}{y}' &= f(x,\underset{\sim}{y}) \quad a \leq x \leq b \;, \\ \underset{\sim}{y}(a) &= \underset{\sim}{y}_0 \;. \end{aligned}\right\} \tag{3.1}$$

We first consider a routine which, given details of the problem, provides only the value y(b) or an indication that the solution cannot be obtained.

The actual parameters of the integration routine must first of all specify this problem.

a) So we need five parameters:

N, the number of equations (not needed in PL/1 or Algol 68)

FCN, a subroutine for evaluating the right-hand sides of the differential equations,

X the initial value a of x,

Y a vector of initial values $\underset{\sim}{y}_0$ of $\underset{\sim}{y}$,

XEND the final value b of x. (3.2)

The final value b of x and the approximation to y(b) will usually be left in X and Y in exit.

(b) Next it must specify in some way the accuracy that we require in the solution. There are several ways in which this can be done. At the simplest level one can specify one real parameter:

EPS	the absolute (or relative) accuracy required in each component of y.	(3.3a)

A more sophisticated routine could require two vectors ,

ABSACC RELACC	The absolute and relative accuracy tolerences in each of the m components of y.	(3.3b)

In this case the routine would attempt to make the ith component of the estimated error ϵ_i satisfy

$$|\epsilon_i| \leq ABSACC_i + RELACC_i\, |y_i|. \qquad (3.4)$$

In practice some compromise between(3.3a) and(3.3b) would be used.

Relative error should not be used for a component for which y_i can be zero. It is often best to use relative accuracy for components that increase and become very large, and to use absolute accuracy otherwise.

(c) Helpful Parameters Our aim is to produce a fully automatic procedure, and the problem is fully specified by the information provided in (a). The user however is likely to know much more about the problem than can be easily obtained from the problem specification. The most useful such information to provide is some idea of the scale of the problem. If a system of equations can have an oscillatory solution, how many cycles are there between initial and final point? If there are 10,000 cycles and the integrator uses a standard method, of the order of 50,000 steps will need to be taken in order to achieve much accuracy; but if there are only a few oscillations it is possible that a correspondingly small number of steps will be sufficient.

Many integrators therefore have one parameter that gives some idea of the scale of the problem. Users may be asked to supply an estimate of the initial steplength to be tried or the maximum step-length to be used. The disadvantage of this sort of estimate is that suitable values are very dependent on the integration method as well as the problem. Because of this Hull and Enwright (1974) suggest that a rough bound SCALE, on the Lipschitz constant of f be used. The larger this constant, the more quickly does the solution change, so the smaller the estimated maximum steplength will be. Hence we would take

maxsteplength := const/SCALE.

The constant here will be very dependent on the method used. Methods will often also need a minimum step length. This is sometimes demanded as a parameter and sometimes calculated within the routine.

(d) A simple routine of this sort is usually called within a loop, and at the rth call the solution is advanced from $a + (r-1)H$ to $a + rH$ say and the solution at $a + rH$ is made available to the calling routine. At the end of the rth call the routine will have available an estimate of the steplength h needed for the integration at the end of the range $[a+(r-1)H, a+rH]$. This will also be a good estimate for the beginning of the range $[a + rH, a + (r+1)H]$, and so, for efficiency, this estimate should be available the second time the routine is called.

A routine using a multistep method will also either have to pass values of the solution at the previous k steps from one call to the next, or repeat its starting procedure at each call - a grossly inefficient thing to do.

There is thus a need to preserve the values of some working quantities from one call of the routine to the next. In all languages this can be done by declaring variables in the calling routine and passing them to the integration routine through the parameter mechanism.

In ALGOL 60 'own' variables of the integrator can be used. In FORTRAN one may be able to rely on local variables of the integration routine being unchanged from one call to the next, but this is not possible in standard FORTRAN and here COMMON must be used. In ALGOL 68 the position of the working space must be passed as a parameter, but since complicated structures can be declared only <u>one</u> parameter is needed whereas in the other languages it may be necessary to use several parameters.

(f) The traditional way to indicate an error condition is to include an integer variable

$$\text{IFAIL}, \tag{3.5}$$

in the parameter list whose value on exit indicates why the routine was left. i.e. IFAIL = 0 if the expected result was obtained, but IFAIL = 1 if the desired accuracy could not be achieved. It is essential therefore to test the value of IFAIL on exit.

In the NAG Library the value of IFAIL when the routine is called is used to specify the failure action. If IFAIL = 1 at the call, the action is as above, but if IFAIL = 0, at the call, and the integrator does not finish normally (i.e. does not finish with IFAIL = 0) a 'hard' failure is said to occur, and a routine is called which prints a message giving the failure number and name of the routine in which the failure occurred and then the run is terminated. (In the ALGOL 68 version of the library instead of including an extra integer in the call a routine name is included; for the 'hard' failure case above, this name is naghard).

4. MORE SOPHISTICATED PROGRAMS

The organisation described in the previous sections is the one traditionally used, but many variants are possible. I want to discuss first the output of intermediate results. If the previously described routines are used, and results at a small interval are needed, the integration routine must be called repeatedly to integrate over a short range. Another possibility is to have the routine integrate over the whole required range, but to call a subroutine specified in the parameter list, to output results. This subroutine could well be supplied with a value of x and the corresponding approximation to y. It could return to the integration routine the value of x at which the next solution is required. Alternatively, the solutions can be supplied at a constant spacing H, where H is a parameter of the integrator.

In a one-step method there seems to be no other way to calculate the solution at specified points than to make sure that such points are mesh points for the normal calculation of the solution. If a multistep method is used however, the most economical way is to interpolate to the required point, making use of the stored information about the solution. This is especially easy in the Adams methods.

The efficient provision of output of solution values at a fine interval might be especially important if graphical output is required, although even there it could be better to output at a fairly wide interval and use the spline interpolation routines usually provided in the graphics package to produce a smooth curve. Due to the limited accuracy of graphical output no more than about three decimal place accuracy will be needed.

An experienced user may sometimes wish to modify the action of the standard integrator to make it more suitable for his particular application. Hull (1974) using as example a routine based on the Runge-Kutta method, suggests no less than eight places where the user might want to take alternative action, and provides for an interrupt at each of these points, so that if a marker is set the integrator will be left at its interrupt point and that particular part of the calculation can be done in the user's routine. One such interrupt enables the user to specify his own maximum step length, bypassing its calculation from SCALE; another enables the user to check and possibly override the step length to be used in the calculation.

The action if an interrupt takes place is as though a special subroutine had been called, but in FORTRAN this is not the best way to implement the facility.

One sometimes needs to locate the value of x at which either one component of the solution is zero, or a function of the solution is zero. If a multistep method is being used this is best implemented in conjunction with the interpolation method of finding the solution at additional points (see Shampine and Gordon (1975)). If a one-step method is used the most efficient way may be to make a change of variable when within one step of the root, so that the function whose zero is required is the new independent variable. One step then locates the required point. This is described in a short note by Wheeler (1959). It is very simple to do if the derivative of the function can be easily calculated.

5. ROUNDING ERRORS

Rounding errors occur of course, in every implementation of every algorithm, and on the whole one must hope that they will be so small that the desired accuracy is achieved; if not one must work to a higher precision. But I do wish to discuss, briefly, two places where one has to pay particular attention to such errors.

The first of these is where an equation is being solved by iteration, as is the case in all implicit methods. One will judge the convergence of the iteration by the closeness of successive iterates and will not be able to reduce this to zero because of rounding errors. One will thus need a constant which depends on the rounding error. The choice of such constants requires considerable care. In more complicated situations more than one such constant may be needed, (e.g. one for an inner and one for an outer iteration); so the interaction between the two will have to be taken into consideration and the routine may need careful experimental tuning - that is the constant may have to be adjusted experimentally, and fine adjustment may have a considerable effect on the efficiency of the method.

Next the precision of the arithmetic, or wordlength of the machine affects the choice of maximum order and step length to be used. It is clearly not worth calculating terms which are of the order of the rounding errors in the y's.

Hull and Enright (1974) use a rather similar criterion for choosing a minimum step length. They chose HMIN so that the estimated truncation error is at least 10 times the round-off error.

So that HMIN can be found before the main calculation starts, they have the user supply YMAX, a bound on the size of the maximum component of y, as part of the calling sequence.

In his programs, Krogh attempts to locate the onset of rounding error on his differences by looking automatically for the order of differences where the rounding error affects their value significantly. Such differences are not used in the calculation.

This enables Krogh to detect and inform the user, when it is impossible to achieve the specified accuracy.

5

RUNGE-KUTTA METHODS

1. INTRODUCTION

Runge-Kutta methods, because of their self-starting property, have a unique place amongst the classical types of methods. We will, in this chapter, discuss some of their properties and also look at some of the limitations that are intrinsic to these methods.

As higher and higher numbers of stages are introduced, storage would become a difficulty for large problems solved by these methods unless it were for the observation that very large problems often have large Lipschitz constants. This generally makes high order Runge-Kutta methods unsuitable for such problems but in any case other methods are usually more efficient and would invariably be used instead.

Thus, we will ignore any question of optimizing with respect to storage requirements and instead focus our attention on accuracy and efficiency for the sort of problems for which Runge-Kutta methods are suitable.

2. DERIVATION OF PARTICULAR METHODS

In Chapter 10, the order conditions for implicit Runge-Kutta methods will be stated. Since explicit Runge-Kutta methods are a special case we can immediately write down their order conditions.

For the method

$$\begin{array}{c|cccc}
0 & & & & \\
c_2 & a_{21} & & & \\
c_3 & a_{31} & a_{32} & & \\
\vdots & \vdots & \vdots & \ddots & \\
c_s & a_{s1} & a_{s2} \cdots & a_{s,s-1} & \\
\hline
 & b_1 & b_2 \cdots & b_{s-1} & b_s
\end{array}$$

(where the blanks represent zeros) the conditions for order p are, using the notation of Chapter 10, that for each rooted tree t with no more than p nodes, the equation

$$\Phi(t) = 1/\gamma(t) \tag{2.1}$$

is satisfied.

Assuming that $s = 4$, these conditions for order 4, together with diagrams of the corresponding trees, are

$$b_1 + b_2 + b_3 + b_4 = 1 \tag{2.2}$$

$$b_2c_2 + b_3c_3 + b_4c_4 = \frac{1}{2} \tag{2.3}$$

$$b_2c_2^2 + b_3c_3^2 + b_4c_4^2 = \frac{1}{3} \tag{2.4}$$

$$b_3a_{32}c_2 + b_4a_{42}c_2 + b_4a_{43}c_3 = \frac{1}{6} \tag{2.5}$$

$$b_2c_2^3 + b_3c_3^3 + b_4c_4^3 = \frac{1}{4} \tag{2.6}$$

$$b_3c_3a_{32}c_2 + b_4c_4a_{42}c_2 + b_4c_4a_{43}c_3 = \frac{1}{8} \tag{2.7}$$

$$b_3a_{32}c_2^2 + b_4a_{42}c_2^2 + b_4a_{43}c_3^2 = \frac{1}{12} \tag{2.8}$$

$$b_4a_{43}a_{32}c_2 = \frac{1}{24} \tag{2.9}$$

Note that for lower p, we take an appropriate subset of these conditions and for lower s, we retain only certain of the terms.

From (2.9) we see that 4 stages are actually necessary since if there were fewer, the only term on the left of this equation would be omitted. As will be shown in Chapter 10, it is generally true that explicit methods must have $s > p$. In fact, the minimum value of s, for given p, is shown in the next table as far is it is known.

p	1	2	3	4	5	6	7
s	1	2	3	4	6	7	9

The general classes of methods with these s and p values are easy to find in the cases $p = 1$, 2 and 3.

For $s = p = 1$:

$$\begin{array}{c|c} 0 & \\ \hline & 1 \end{array}$$

This is the (unique) method of Euler.

For $s = p = 2$:

$$\begin{array}{c|cc} 0 & & \\ c & c & \\ \hline & 1-\frac{1}{2c} & \frac{1}{2c} \end{array}$$

This one parameter family is of the required order for any (non-vanishing) c.

For $s = p = 3$ there are three families of which the first two are

$$\begin{array}{c|ccc} 0 & & & \\ \frac{2}{3} & \frac{2}{3} & & \\ 0 & -\frac{1}{4b} & \frac{1}{4b} & \\ \hline & \frac{1}{4}-b & \frac{3}{4} & b \end{array} \qquad \begin{array}{c|ccc} 0 & & & \\ \frac{2}{3} & \frac{2}{3} & & \\ \frac{2}{3} & \frac{2}{3}-\frac{1}{4b} & \frac{1}{4b} & \\ \hline & \frac{1}{4} & \frac{3}{4}-b & b \end{array}$$

Each of these is a one parameter family with $b \neq 0$. The third family has c_2, c_3 as parameters with $c_2 c_3 (c_2 - \frac{2}{3})(c_3 - c_2) \neq 0$.

The derivation of methods with $s = p = 4$ is more complicated but a simplification results if

$$\sum_i b_i a_{ij} = b_j(1 - c_j), \qquad j = 1, 2, 3, 4 \tag{2.10}$$

(which implies that $c_4 = 1$) since this allows (2.3), (2.5), (2.8) and (2.9) to be omitted from the set of order conditions. It is also interesting that (2.10) is a necessary consequence of (2.2) to (2.9).

We now outline the steps that are taken to derive a particular method of this order. These steps can be carried out on the proviso that inconsistent linear systems do not arise

Step 1: Select values of c_2, c_3 and set $c_4 = 1$.

Step 2: Solve for b_1, b_2, b_3, b_4 from (2.2), (2.3), (2.4) and (2.6).

Step 3: Solve for a_{32} from $b_3(1 - c_3)\, a_{32} c_2 = \frac{1}{24}$ (the difference of (2.5) and (2.7)).

Step 4: Solve for a_{42}, a_{43} from (2.10).

Step 5: Calculate $a_{21} = c_2$, $a_{31} = c_3 - a_{32}$, $a_{41} = 1 - a_{42} - a_{43}$.

In the case $c_2 = c_3 = \frac{1}{2}$, step (2) leads to a choice of b_2 and b_3

under the conditions $b_2 + b_3 = \frac{2}{3}$, $b_3 \neq 0$. We have in particular the well known method

$$
\begin{array}{c|cccc}
0 & & & & \\
\frac{1}{2} & \frac{1}{2} & & & \\
\frac{1}{2} & 0 & \frac{1}{2} & & \\
1 & 0 & 0 & 1 & \\
\hline
 & \frac{1}{6} & \frac{1}{3} & \frac{1}{3} & \frac{1}{6}
\end{array}
$$

Using an extension of these techniques, methods can easily be derived of order 5 with 6 stages. For example, for any λ and non-zero μ we have the method

$$
\begin{array}{c|cccccc}
0 & & & & & & \\
\mu & \mu & & & & & \\
\frac{1}{4} & \frac{1}{4} - \frac{1}{32\mu} & \frac{1}{32\mu} & & & & \\
\frac{1}{2} & 4\lambda\mu - \lambda & \lambda & \frac{1}{2} - 4\lambda\mu & & & \\
\frac{3}{4} & -\frac{3}{16} + \frac{3}{32\mu} + \frac{3\lambda}{4} & -3\lambda\mu - \frac{3}{32\mu} - \frac{3\lambda}{4} & \frac{3}{8} + 3\lambda\mu & \frac{9}{16} & & \\
1 & \frac{5}{7} - \frac{12\lambda}{7} - \frac{2}{7\mu} + \frac{48\lambda\mu}{7} & \frac{2}{7\mu} + \frac{12\lambda}{7} & \frac{6}{7} - \frac{48\lambda\mu}{7} & -\frac{12}{7} & \frac{8}{7} & \\
\hline
 & \frac{7}{90} & 0 & \frac{32}{90} & \frac{12}{90} & \frac{32}{90} & \frac{7}{90}
\end{array}
$$

The derivation of order 6 methods is also straightforward but orders 7 and 8 are more complicated. All known methods of order 8 have 11 stages and it is not know whether methods exist of this order with 10 stages.

3. METHODS WITH ERROR ESTIMATES

An efficient way of computing local estimates is to use an $s + 1$ stage $p + 1$ order method

$$\begin{array}{c|ccccc}
0 & & & & & \\
c_2 & a_{21} & & & & \\
c_3 & a_{31} & a_{32} & \ddots & & \\
\vdots & \vdots & \vdots & & \ddots & \\
c_{s+1} & a_{s+1,1} & a_{s+1,2} & \cdots & a_{s+1,s} & \\
\hline
 & b_1 & b_2 & \cdots & b_s & b_{s+1}
\end{array}$$

such that the s stage method

$$\begin{array}{c|ccccc}
0 & & & & & \\
c_2 & a_{21} & & & & \\
c_3 & a_{31} & a_{32} & \ddots & & \\
\vdots & \vdots & \vdots & & \ddots & \\
c_s & a_{s1} & a_{s2} & \cdots & a_{s,s-1} & \\
\hline
 & a_{s+1,1} & a_{s+1,2} & \cdots & a_{s+1,s-1} & a_{s+1,s}
\end{array}$$

embedded within it, is of order p. If the embedded method is used for the computation of y_n from y_{n-1} then the $s + 1$ stage method can be used as an error estimate. In effect, this procedure uses only s derivative calculations per step, since the evaluation of $f(x_n, y_n)$ is required for the following step in any case.

If we define $\Phi(t)$ for this scheme in the usual way outlined in Chapter 10 but with b_{s+1} replaced by zero and if we write β for the value of b_{s+1}, then the order conditions for the $s + 1$ stage method, taking account of the order requirements of the embedded method, are

$$\Phi(t) = \frac{1}{\gamma(t)}(1 - \beta r(t)) \qquad (3.1)$$

for all trees t such that $r(t) \leq p + 1$. If $\beta \neq 0$ then $a_{s+1,1}$, $a_{s+1,2}, \ldots, a_{s+1,s}$ (which do not appear in (3.1)) can be computed from

$$a_{s+1,i} = (b_i(1 - c_i) - \sum_j b_j a_{ji})/\beta .$$

Although this type of method is more complicated, in terms of its detailed derivation, than the usual type of Runge-Kutta method we give the examples

$s = p = 1$

$$\begin{array}{c|cc} 0 & & \\ 1 & 1 & \\ \hline & \frac{1}{2} & \frac{1}{2} \end{array}$$

$s = 3$, $p = 2$ (A re-interpretation of the classical 4th order method)

$$\begin{array}{c|cccc} 0 & & & & \\ \frac{1}{2} & \frac{1}{2} & & & \\ \frac{1}{2} & 0 & \frac{1}{2} & & \\ 1 & 0 & 0 & 1 & \\ \hline & \frac{1}{6} & \frac{1}{3} & \frac{1}{3} & \frac{1}{6} \end{array}$$

$s = 4$, $p = 3$

$$\begin{array}{c|ccccc} 0 & & & & & \\ \frac{2}{7} & \frac{2}{7} & & & & \\ \frac{4}{7} & -\frac{8}{35} & \frac{4}{5} & & & \\ \frac{6}{7} & \frac{29}{42} & -\frac{2}{3} & \frac{5}{6} & & \\ 1 & \frac{1}{6} & \frac{1}{6} & \frac{5}{12} & \frac{1}{4} & \\ \hline & \frac{11}{96} & \frac{7}{24} & \frac{35}{96} & \frac{7}{48} & \frac{1}{12} \end{array} .$$

Methods of higher orders have been given by Fehlberg (1968) but these have $\beta = 0$ in our notation and suffer from a disadvantage that the two constituent Runge-Kutta methods are based on the same quadrature formula. This leads to poor step-size control for some problems.

4. LOCAL TRUNCATION ERROR

In the performance of a single step by a Runge-Kutta method, the local truncation error can be found in the form of an asymptotic expansion by taking the difference of (2.5) and (2.9) in Chapter 10, where it is supposed that $y(x_{n-1}) = y_{n-1}$. We have, for a method of order p,

$$y_n - y(x_n) = \sum_{r(t)>p} \frac{h^{r(t)}}{\sigma(t)} (\Phi(t) - \frac{1}{\gamma(t)})F(t)(y_{n-1}). \tag{4.1}$$

If we have bounds on the various partial derivatives occuring in this expansion, the error can be estimated. Usually, however, such information is not easy to obtain.

Let y_n be computed by the methods of the previous section and let $\tilde{y}_n$ be the additional approximation of order $p + 1$. From (4.1) we have

$$y_n = y(x_n) + \sum_{r(t)=p+1} \frac{h^{p+1}}{\sigma(t)} (\phi(t) - \frac{1}{\gamma(t)}) F(t)(y_{n-1}) + O(h^{p+2})$$

$$\tilde{y}_n = y(x_n) + O(h^{p+2})$$

Thus the local error is $y_n - \tilde{y}_n + O(h^{p+2})$ and can normally be approximated accurately by $y_n - \tilde{y}_n$, at least for sufficiently small $|h|$.

This error approximation can be used for two purposes (a) as a criterion that the error is small enough in the present step (b) as a means of selecting a suitable step-size for the succeeding step. If we are going to control local errors in this way we should examine the relationship between local and global errors.

If a method produces errors at the rate (per distance along the trajectory) $e(x)$ as x varies from $x = a$ to $x = b$, then the global error will be approximately $\int_a^b G_x(b)e(x)dx$ where for each $c \in [a, b]$, G_c is the solution to the matrix differential equation

$$G_c'(x) = f'(y(x))G_c(x)$$

with initial value

$$G_c(c) = I .$$

If, for example, f' is a constant matrix M, then $G_c(x)$ is $\exp((x-c)M)$.

5. OPTIMAL CHOICE OF STEP-SIZE

Suppose that in integrating from x_{n-1} to x_n with step size h_n

the error is approximately $v(x_{n-1})h_n^{p+1}$. Since this corresponds to a rate of error production of about $v(x_{n-1})h_n^p$ we write $e(x) = v(x)h(x)^p$ where $h(x)$ denotes the value of a step selection function. Write $\phi(x) = \|G_x(b)v(x)\|$ and we obtain a bound on the integral that approximates the total error

$$E = \int_a^b \phi(x)\ h(x)^p\, dx .$$

On the other hand, the cost will be proportional to the number of steps which is approximately

$$C = \int_a^b h(x)^{-1} dx .$$

If we wish to keep E down to some fixed value but minimize C, a calculus of variations argument tells us that $\phi(x)\ h(x)^{p-1}/h(x)^{-2}$ should be kept constant. That is, the contribution to the final error should be the same for each step.

Although it is not usually possible to determine this optimal choice of h a priori, this result is a useful guide in designing good programs.

6. OPTIMAL CHOICE OF ORDER

The choice between two possible methods will depend on their actual costs when used to obtain approximately equal accuracy for the same computation. Since the choice of best method will vary along the solution trajectory, a flexible program will switch between methods of different orders selecting the one that gives the best performance locally. In the case of Runge-Kutta methods this would require the occassional use of exploratory steps with alternative methods to the one currently in use.

Consider two methods which, with step sizes h_1, h_2 respectively, generate errors at the rates $v_1(x)h^{p_1}$, $v_2(x)h^{p_2}$ and accumulate costs at the rates c_1/h_1, c_2/h_2. Let h denote h_1/c_1 or h_2/c_2 so that integrating with either of the methods will cost the same for a single step. We then find that the two methods are producing errors at the rates $\|v_1(x)\|\, c_1^{p_1}\, h^{p_1}$, $\|v_2(x)\|\, c^{p_2}\, h^{p_2}$. If $p_1 < p_2$ the first will be the less for large h and the second for small h. To find the break-even point, let ε be the tolerable error per distance integrated when the two methods are equally efficient. We find

$$\varepsilon^{(1/p_1 - 1/p_2)} = \|v_1(x)\|^{1/p_1} \|v_2(x)\|^{-1/p_2} c_1 c_2^{-1}$$

To use this formula, $\|v_1(x)\|$, $\|v_2(x)\|$ are estimated using local error estimates as we have already described. Further information on method and step selection is to be found in Gear (1971b).

7. ASSESSMENTS OF METHODS

The analysis of the previous section was approximate in several respects. In particular, the asymptotic expansions of errors were assumed to be close enough for comparisons of methods even when large step sizes are used. This incorrect comparison is to the disadvantage of high order methods and many programs compensate for it by introducing a bias towards high orders.

Another complicating factor is that of stability. Even for non-stiff problems, there is a threshold step-size above which the qualitative behaviour of the computed solution changes and it becomes completely unsatisfactory. For example, in the case of linear problems this will happen when $h\lambda$ (for λ some eigenvalue of the Jacobian matrix for the problem) moves out of the stability region. Thus the ability of a method to yield good results depends on its stability characteristics as well as on its order. If $s = p$ (which is possible for $p = 1, 2, 3, 4$) there is no choice of the stability region but if s is made greater than p, the extra flexibility can be used to extend the stability region.

Although we have not looked at the question of rounding errors, it is clear that the accumulation of such errors within a step will depend in some way on the coefficients of the method. Without carrying out a detailed analysis of this, it is still possible to say that negative signs appearing amongst the coefficients of the method, especially $b_1, b_2, \ldots, b_s$ is a warning that there is a danger of trouble. Many high order methods where negative signs occur have large values for some $|a_{ij}|$ and this will lead to loss of accuracy through cancellation of significant digits.

Finally, a good program which uses Runge-Kutta methods should have available formulae for a wide range of orders, say from 2 to 5. As has already been indicated, low order formula are often more efficient when there are undemanding accuracy requirements and the opposite is certainly the case. Also, of course, low order methods should be available for problems which have discontinuities in their low order derivatives so that the asymptotic expansion of error is not valid.

6

IMPLEMENTATION OF LINEAR MULTISTEP METHODS

1. INTRODUCTION

This chapter deals with the implementation of linear multistep methods. Although we are not concerned with stiff problems, much of the discussion is relevant to the implementation of multistep methods for stiff problems, Chapter 11. Certain aspects of this topic are discussed thoroughly in Gear (1971b) and a new book by Shampine and Gordon (1975) contains a full treatment of the implementation of Adams method. The work of Krogh (e.g. 1969a, 1969b, 1974) is of major importance in this area. Over the last seven or eight years several good variable-step, variable order programs based on the Adams methods have been produced, by the above mentioned authors and also by Sedgwick (1973) and by the NAG library, among others. These programs differ, in some cases in important respects. Their importance and advantages will be stressed in Chapter 8, in comparisons with other methods. Our present task is to describe such programs by considering the major considerations involved in their construction. One can spend a great deal of time in developing such programs and it will not be possible to go into all the details connected with making the program reliable and robust. For a good idea of what is involved, see Krogh (1974).

For the most part it will be convenient to consider the problem of a single first order equation,

$$y' = f(x, y), \; y(a) = \eta \qquad (1.1)$$

with the solution required at certain values of x in the range [a, b]. The extension to first order systems, which is mostly straightforward, will be pointed out where necessary.

In §2 we define the Adams formulae and explain why they are chosen to form the basis for the program. Then in §3 various algorithms for step-by-step computation are considered. The problem of changing stepsize and its dependence on the representation of the formulae, together with local error estimation is discussed in §4. Briefly, in §5, we indicate generalisations to higher order equations and implicit differential equations. The topic is continued in §6 with a discussion of the implementation of variable order and is concluded with a comparison of the major alternatives, as a whole, on theoretical and practical grounds, in §7.

2. WHY THE ADAMS METHODS?

All the major implementations of variable order multistep methods have utilised the Adams class of methods. We repeat the definition here for convenience. Integrate (1.1) to obtain,

$$y(x_{n+1}) = y(x_n) + \int_{x_n}^{x_{n+1}} f(x,\ y(x))dx. \tag{2.1}$$

Denote the minimum degree polynomial that interpolates the values f_n, $f_{n-1},\ldots,\ f_{n-k+1}$ by $P_{k,n}(x)$, where $f_r = f(x_r,\ y_r)$. Then the modern usage is to refer to the formula

$$y_{n+1} = y_n + \int_{x_n}^{x_{n+1}} P_{k,n}(x)dx \tag{2.2}$$

as the k-step Adams-Bashforth (explicit) formula. In §4 we examine particular representations of this formula, suitable for computation. Similarly,

$$y_{n+1} = y_n + \int_{x_n}^{x_{n+1}} P_{k+1,n+1}(x)dx \tag{2.3}$$

is called the k-step Adams-Moulton (implicit) formula. The formulae (2.2) and (2.3) are of orders k and $k+1$ respectively.

The Adams formulae possess the important property of being strongly stable. In Chapter 2 it is shown that this property is obtained by making the extraneous roots of the associated polynomial $\rho(s)$, strictly less than unity. The Adams formulae can be characterized as being maximal order formulae with the property that these extraneous roots are all zero. This ensures an adequate region of absolute stability for most problems. (Other multistep methods need to be considered in an implementation designed for stiff problems, Chapter 11.)

It is possible to enlarge the region of absolute stability in a simple way by taking a linear combination of the Adams-Moulton correctors of orders k and $k+1$. When the size of the region of absolute stability is limiting the stepsize (rather than the accuracy requirement) the tendency, provided the order selection process is well designed(!), is for the algorithm to gravitate towards low order (small k). The reason for this is that the regions of stability get rapidly smaller as k is increased. To cope better with problems which present such stability difficulties it is probably worth trying to improve the stability of the algorithms at low order. For example if we use (2.2) and (2.3) with $k=1$ in a PECE algorithm it may be written, where $h = x_{n+1} - x_n$,

$$y_{n+1}^{p} = y_n + hf_n \,, \tag{2.4}$$

$$y_{n+1} = y_{n+1}^{p} + \frac{h}{2}(f(x_{n+1}, y_{n+1}^{p}) - f_n) \,, \tag{2.5}$$

$$f_{n+1} = f(x_{n+1}, y_{n+1}) \,. \tag{2.6}$$

This is the Euler predictor and the Trapezium rule corrector. If we replace (2.5) by

$$y_{n+1} = y_{n+1}^{p} + \frac{wh}{2}\,(f(x_{n+1}, y_{n+1}^{p}) - f_n) \tag{2.7}$$

we have a PECE method which is absolutely stable for $y' = \lambda y$ if,

$$\left|1 + h\lambda + \frac{wh^2\lambda^2}{2}\right| < 1 \,. \tag{2.8}$$

The formula (2.7) is a linear combination of the Trapezium rule and the backward Euler first order formula. Using (2.5), $w = 1$, the algorithm is stable for $-2 < h\lambda < 0$. Krogh (1974) suggests the value $w = \frac{1}{4}$ for which the corresponding interval is $-8 < h\lambda < 0$. To decide an optimum value it is necessary to plot the stability region in the complex $h\lambda$ - plane.

A similar investigation could be done for other values of k. It is not worthwhile for large values because when such values are employed by the program, it is almost certainly accuracy rather than stability that is limiting the stepsize. Of course, if stability becomes a very severe restriction on the stepsize, a switch to methods designed for stiff problems is required.

3. STEP-BY-STEP ALGORITHMS

In Chapter 1 predictor-corrector modes with a fixed number of corrections and the algorithm of iterating the corrector to convergence were presented. Hull and Creemer (1963) argue that the most efficient algorithms, in general, are those that employ two or three corrections. We have to decide whether to use a fixed algorithm, usually PECE or PECEC, using two derivative evaluations on each step, or the mode of iteration to convergence with a limit on the number of corrections allowed. PEC should also be considered.

Gear (1969b) uses iteration to convergence. The strategy employed permits at most three corrections; if convergence is not achieved then the stepsize is reduced and the step repeated. His strategy for step-size selection and the fact that the explicit methods provide accurate starting values ensure that convergence is almost always achieved. This is a sound algorithm.

Some theorems on the region of absolute stability associated with the various ways of using Adams methods are given in Hall (1974), for the case of constant stepsize. The following discussion is based on these results.

Let P_k and C_k denote the explicit and implicit Adams methods respectively, of order k (note that (2.3) and (2.4) define P_k and C_{k+1}). The algorithm $P_kEC_{k+1}E$ has better stability properties than the iterated corrector for $k > 5$. For $k \leqslant 5$ the difference in stability is not very significant. (At very low orders the effective region of stability for the correctors is reduced by the requirement that the iteration must converge). As far as stability is concerned it is, on balance, difficult to choose between $P_kEC_{k+1}E$ and the iterated corrector, but on cost the former gets the preference. The local error estimate can be computed before the second evaluation. Other possibilities, e.g. P_kEC_kE, have comparable stability regions; at high order they are slightly better. $P_kEC_{k+1}E$ is still to be preferred on balance. It has better stability at low order, where it matters, and potentially better accuracy at high order. This is the algorithm preferred by Krogh (1974) on the basis of computing stability regions.

Algorithms such as P_kEC_k or P_kEC_{k+1} have the advantage of requiring only one evaluation per step. However, this is more than offset for most computations by the fact that the regions of stability are very poor compared with any two evaluation algorithm, for all values of k. It is still worth considering their use in situations where derivative evaluations are expensive and where it is clear that the accuracy requirement will be limiting the stepsize. Their use is sometimes advocated for computations associated with the N-body problem. Ideally it should be made possible for the user of the program to select between $P_kEC_{k+1}E$ and P_kEC_{k+1}.

4. REPRESENTATION, STEP CHANGING, ERROR ESTIMATION.

In a general way, an efficient implementation of the Adams methods depends on the extent to which information about the behaviour of the solution can be extracted from the back values $f_n, f_{n-1}, \ldots, f_{n-k+1}$. This information is essential for the correct selection of stepsize and order. A discontinuity in a higher derivative of the solution, requiring a reduction in order, might not be apparent from the above values and yet could be detected in some of the representations we consider.

Suppose first that we intend to use the constant stepsize form of the Adams formulae. Then an attractive possibility is to store the backward differences

$$\nabla^i f_n, \; i = 0, 1, \ldots, k-1,$$

defined by

$$\nabla^0 f_n = f_n ,$$

$$\nabla^i f_n = \nabla^{i-1} f_n - \nabla^{i-1} f_{n-1} .$$

In this case, the predictor (2.2) takes the form

$$y_{n+1}^p = y_n + h \sum_{i=0}^{k-1} \gamma_i \nabla^i f_n , \tag{4.1}$$

where the constants γ_i are independent of k.

In a $P_k EC_{k+1} E$ algorithm the corrector (2.3) can be written

$$y_{n+1} = y_n + h \sum_{i=0}^{k} \gamma_i^* \nabla_p^i f_{n+1} , \tag{4.2}$$

where ∇_p^i denotes the ith backward difference using $f(x_{n+1}, y_{n+1}^p)$ for f_{n+1}. We can also show (Henrici 1961) that

$$\gamma_0^* = \gamma_0 = 1, \quad \sum_{i=0}^{k} \gamma_i^* = \gamma_k ,$$

and it follows that (4.2) may be simplified for computation to

$$y_{n+1} = y_{n+1}^p + h \gamma_k \nabla_p^k f_{n+1} . \tag{4.3}$$

In a $P_k EC_k E$ algorithm simply replace γ_k by γ_{k-1} in (4.3), which corresponds to taking one fewer term in (4.2). The Milne error estimate, Chapter 3, for this algorithm is simply the difference between these two possibilities giving,

$$\text{estimated error} = h\gamma_k^* \nabla_p^k f_{n+1} .$$

The same estimate can be used for the $P_k EC_{k+1} E$ algorithm, when the corrector is potentially more accurate. If the accuracy requirement is satisfied, evaluate $f(x_{n+1}, y_{n+1})$ and form the corrected differences $\nabla^i f_{n+1}$ to complete the step.

If a stepsize change from h to h' is required, prior to the above step, the usual method is to interpolate for the values $\tilde{f}_n$, $\tilde{f}_{n-1}$, ..., $\tilde{f}_{n-k+1}$ at the points x_n, $x_n - h'$,..., $x_n-(k-1)h'$, using the interpolation polynomial $P_{k,n}(x)$ through f_n, f_{n-1},..., f_{n-k+1}. Thus

$$\tilde{f}_{n-i} = P_{k,n}(x_n - i\,h'), \quad i = 0, 1,\ldots, k-1. \tag{4.4}$$

Efficient algorithms for this in terms of backward differences are given by Krogh (1973a) in the cases where the stepsize is doubled or halved. The cost is roughly comparable with the work involved in carrying out one step of the method, except that no derivative evaluations are required. Stepsize changes by a factor other than 2 or $\frac{1}{2}$ would be possible but much less convenient.

An attractive alternative possibility to the above is that based on the Nordsieck (1962) formulation which is used by Gear (1969, 1971b). In this case the back values are stored in the form

$$y_n,\ hy'_n,\ldots,\ h^k y_n^{(k)}/k! \tag{4.5}$$

In this notation the scaled derivatives are actually those of the interpolation polynomial to f_n, f_{n-1},..., f_{n-k+1}. More precisely

$$y_n^{(i)} = \left.\frac{d^{i-1}P_{k,n}(x)}{dx^{i-1}}\right|_{x=x_n} . \tag{4.6}$$

This form is described in detail in Gear (1971b). Nordsieck's motivation was to simplify the above interpolation process (4.4) for changing the stepsize. In fact an arbitrary change is achieved simply by multiplying the ith scaled derivative in (4.5) by the factor α^i where $\alpha = h'/h$.

Of the two the Nordsieck form is therefore more flexible and they require a comparable amount of computation per step. The backward differences are conceptually simpler. If the order is reduced by one by omitting the last term in (4.1) then we obtain the explicit Adams formula or order $(k-1)$. The natural way, when using the Nordsieck vector of information (4.5), of reducing order is to omit the last term. Thereafter we are not using an Adams formula. For example when $k = 2$

the Nordsieck vector of saved information, (4.5), is y_n, hfn, $\frac{h}{4}(3f_n - 4f_{n-1} + f_{n-2})$, $\frac{h}{3}(f_n - 2f_{n-1} + f_{n-2})$. The predicted value y_{n+1}^p is the sum of those terms. If the last term is omitted we do not discard f_{n-2}. The formula would no longer be an Adams formula.

Some further discussion of these forms under stepsize changes appears in §7. We conclude this section by presenting a variable step formulation which is always a faithful representation of the Adams methods under stepsize or order variation. It is more costly per step but the discussion in §7 will suggest that it may be the best choice, taking reliability and cost into account. Let $f_{n,n-1,\ldots,n-i}$ denote the ith divided difference of f defined recursively by

$$f_{n,n-1,\ldots,n-i} = \frac{f_{n,n-1,\ldots,n-i+1} - f_{n-1,n-2,\ldots,n-i}}{x_n - x_{n-i}} . \tag{4.7}$$

The interpolation polynomial in terms of divided differences takes the form

$$P_{k,n}(x)=f_n+(x-x_n)f_{n,n-1}+\ldots+(x-x_n)\ldots(x-x_{n-k+2})f_{n,n-1,\ldots,n-k+1} , \tag{4.8}$$

$$P_{k+1,n+1}(x) = P_{k,n}(x) + (x-x_n)\ldots(x-x_{n-k+1})f_{n+1,n,\ldots,n-k+1} . \tag{4.9}$$

Let g_{ij} be the j-fold integral

$$g_{ij} = \int_{x_n}^{x_{n+1}} \int_{x_n}^{x} \cdots \int_{x_n}^{x} (x-x_n)(x-x_{n-1})\ldots(x-x_{n-i+1})dx . \tag{4.10}$$

Then the $P_k EC_{k+1} E$ algorithm takes the form,

$$\left.\begin{aligned} y_{n+1}^p &= y_n + \sum_{i=0}^{k-1} g_{i1} f_{n,n-1,\ldots,n-i} , \\ f_{n+1}^p &= f(x_{n+1}, y_{n+1}^p) , \\ y_{n+1} &= y_{n+1}^p + g_{k1} f^p_{n+1,n,\ldots,n-k+1} , \\ f_{n+1} &= f(x_{n+1}, y_{n+1}) . \end{aligned}\right\} \tag{4.11}$$

As usual the notation $f^{p}_{n+1,n,\ldots,n-k+1}$ refers to divided differences formed using f^{p}_{n+1}. If the additional extrapolated value $f^{*}_{n+1} = P_{k,n}(x_{n+1})$ is obtained from

$$f^{*}_{n+1} = \sum_{i=0}^{k-1} g_{i0} f_{n,n-1,\ldots,n-i} \tag{4.12}$$

then from (4.9) we obtain, putting $x = x_{n+1}$,

$$f^{p}_{n+1} = f^{*}_{n+1} + g_{k,0} f^{p}_{n+1,n,\ldots,n-k+1} \quad . \tag{4.13}$$

Hence the corrected value in (4.11) may be written,

$$y_{n+1} = y^{p}_{n+1} + \frac{g_{k1}}{g_{k0}} (f^{p}_{n+1} - f^{*}_{n+1}) \quad . \tag{4.14}$$

To use P_kEC_kE replace k by $k-1$ in (4.14). The difference between these two possibilities may again be used to estimate the error in either case. We find,

$$\text{estimated error} = \frac{1}{g_{k,0}} (g_{k1} - (x_{n+1} - x_{n-k+1}) g_{k-1,1}) (f^{p}_{n+1} - f^{*}_{n+1}) \ . \tag{4.15}$$

The algorithm depends critically on an efficient method for computing the variable coefficients g_{ij}. From (4.10), using integration by parts repeatedly we can show that

$$g_{ij} = (x_{n+1} - x_{n-i+1}) g_{i-1,j} - j g_{i-1,j+1} \ . \tag{4.16}$$

Since $g_{0j} = (x_{n+1} - x_n)^j / j!$, (4.16) can be used to compute the required coefficients in a triangular array, as indicated.

$k = 2$

i \ j	0	1	2	3
0	g_{00}	g_{01}	g_{02}	g_{03}
1	g_{10}	g_{11}	g_{12}	
2	g_{20}	g_{21}		
3	g_{30}			

Each entry is obtained from two elements in the row above as indicated by (4.16). Note that (4.15) can also be simplified using (4.16).

In using this representation it is important to realise that if the stepsize is held constant over s-steps then the coefficients g_{ij}, $i < s$, remain constant. We can take advantage of this fact by keeping the stepsize constant, if our step selection scheme indicates that only a relatively small increase would be possible. However, it is important to note that in the last scheme, the overhead associated with computing the variable coefficients becomes less significant when integrating a large system of equations.

The last development is due in essence to Krogh (1974), although his treatment is for modified divided differences which reduce to simple backward differences when the stepsize is held constant.

5. GENERALISED ALGORITHMS

We indicate briefly in this Section how the Adams methods can be generalised to deal with higher order equations directly. In practice it is quite easy to write the program so that systems of equations of different orders can be integrated directly, without reduction to equivalent first order systems as in Chapter 1. Consider the dth order equation,

$$y^{(d)} = f(x, y, y', \ldots, y^{(d-1)}) . \tag{5.1}$$

Integrating d-s times leads to the identity,

$$y^{(s)}(x_{n+1}) = \sum_{i=0}^{d-1-s} \frac{h^i}{i!} y^{(s+i)}(x_n) + \int_{x_n}^{x_{n+1}} \int_{x_n}^{x} \cdots \int_{x_n}^{x} f(x,y(x),\ldots,y^{(d-1)}(x))dx .$$

Using the interpolation polynomial $P_{k,n}(x)$ as an approximation to f, we obtain the generalised explicit Adams formula

$$y_{n+1}^{(s)} = \sum_{i=0}^{d-1-s} \frac{h^i}{i!} y_n^{(s+i)} + \int_{x_n}^{x_{n+1}} \int_{x_n}^{x} \cdots \int_{x_n}^{x} P_{k,n}(x)\, dx . \tag{5.2}$$

As in §3 we present the $P_kEC_{k+1}E$ algorithm. This could be done in any formulation used in §4; we give that in terms of divided differences which seems likely to become accepted as the most suitable form for using Adams methods. It is

$$y_{n+1}^{(s)p} = \sum_{i=0}^{d-1-s} \frac{h^i}{i!} y_n^{(s+i)} + \sum_{i=0}^{k-1} g_{i,d-s} f_{n,n-1,\ldots,n-i} \quad ,s=0,1,\ldots, d-1,$$

$$f_{n+1}^{p} = f(x_{n+1}, y_{n+1}^{p},\ldots, y_{n+1}^{(d-1)p}) ,$$

$$y_{n+1}^{(s)} = y_{n+1}^{(s)p} + g_{k,d-s} f^{p}_{n+1,n,\ldots, n-k+1}, \quad s = 0,1,\ldots, d-1, \tag{5.3}$$

$$f_{n+1} = f(x_{n+1}, y_{n+1},\ldots, y_{n+1}^{(d-1)}) .$$

Error estimates can be developed exactly as in §4 for the first order equation. Note that the coefficients are exactly those occurring in the definition (4.10). This triangular array has to be extended a little to provide sufficient coefficients for (5.3). This seems an attractive algorithm and further details of implementation may be found in Krogh (1974).

There is as yet little theory to support an argument in favour of reducing higher order equations to first order or treating them directly. A simple computational advantage of (5.3) is that only one set of divided differences need be formed and updated on each step. A reduction to d first order equations would require d sets of differences.

Finally we note that if (5.1) were replaced by a dth order equation that could not be 'solved' for $y^{(d)}$, the algorithm can still be implemented. Any equations in a system of this implicit form would be solved for $y_{n+1}^{(d)}$ from the other predicted values, replacing the simple computations of f^p_{n+1} and f_{n+1}. A Newton type iteration could be used using a numerical approximation to the Jacobian.

6. VARIABLE ORDER, STARTING THE INTEGRATION

It is very easy to vary the order when using multistep methods. The order depends directly on the number of back values stored. After completing any step the order can be increased by one if none of the back values used on the last step are discarded. The order can be decreased to any desired value by discarding the appropriate number of back values. However, the development of a correct strategy for order selection is difficult and interesting. This aspect is of great importance for the performance of the method. The main ideas are discussed in this Section. For simplicity the backward difference formulation of the Adams methods is used, but the main ideas are independent of the formulation.

Suppose that a $P_k EC_{k+1} E$ algorithm is used. In order to decide whether the order should be increased it is necessary to have available one more difference (back value) than is used at the current order.

Given y_n, $\nabla^i f_n$, $i = 0, 1, \ldots, k$, the $(n+1)$st step is described by

$$\left.\begin{aligned}
y^p_{n+1} &= y_n + h \sum_{i=0}^{k-1} \gamma_i \nabla^i f_n \;, \\
f^p_{n+1} &= f(x_{n+1}, y^p_{n+1}) \;, \\
y_{n+1} &= y^p_{n+1} + h\gamma_k \nabla^k_p f_{n+1} \;, \\
E_k &= h\gamma^*_k \nabla^k_p f_{n+1} \;, \\
f_{n+1} &= f(x_{n+1}, y_{n+1}) .
\end{aligned}\right\} \tag{6.1}$$

The second derivative evaluation is delayed until a test is made to see whether the estimated error, E_k, satisfies the local accuracy requirement, which we take to be,

$$|E_k| < \text{TOL}$$

where TOL is supplied by the user. Suppose this test is satisfied and the step is completed by evaluating f_{n+1} and forming the updated differences, defined by

$$\nabla^0 f_{n+1} = f_{n+1},\; \nabla^i f_{n+1} = \nabla^{i-1} f_{n+1} - \nabla^{i-1} f_n,\; i = 1, 2, \ldots \,. \tag{6.2}$$

At this point a decision can be made to keep the same order or to change the order by one, up or down, before the next step. <u>This is done on the</u>

basis of examining what the estimated error would have been if these orders had been used in the computation of y_{n+1}.

First consider a reduction in order. Then the estimated error would have been

$$E_{k-1} = h\gamma^*_{k-1}[\nabla_p^{k-1} f_{n+1} + f(x_{n+1}, y^p_{n+1} - h\gamma_{k-1}\nabla^{k-1} f_n) - f(x_{n+1}, y^p_{n+1})] \qquad (6.3)$$

since a different predictor, of order one less, would have been used. By the mean value theorem we may write,

$$E_{k-1} = h\gamma^*_{k-1}[\nabla_p^{k-1} f_{n+1} - h\gamma_{k-1}\nabla^{k-1} f_n \cdot \frac{\partial f}{\partial y}] \ . \qquad (6.4)$$

In practice to avoid the extra derivative evaluation required for (6.3) the decision to reduce the order is based on the estimate

$$\tilde{E}_{k-1} = h\gamma^*_{k-1}\nabla_p^{k-1} f_{n+1} , \qquad (6.5)$$

which is immediately available and, from (6.4), asymptotically valid.

To consider an increase in order the estimated error is

$$E_{k+1} = h\gamma^*_{k+1}\nabla_p^{k+1} f(x_{n+1}, y^p_{n+1} + h\gamma_k\nabla^k f_n) , \qquad (6.6)$$

where the predicted difference is based on a predictor with one more term included. Comparing with the algorithm (6.1) it is apparent that the corrected differences, (6.2), should be used in estimating E_{k+1}. We obtain

$$E_{k+1} = h\gamma^*_{k+1}[\nabla^{k+1} f_{n+1} + f(x_{n+1}, y^p_{n+1} + h\gamma_k\nabla^k f_n) - f(x_{n+1}, y^p_{n+1} + h\gamma_k\nabla_p^k f_{n+1})] ,$$

leading to

$$E_{k+1} = h\gamma^*_{k+1}[\nabla^{k+1} f_{n+1} - h\gamma_k\nabla_p^{k+1} f_{n+1} \frac{\partial f}{\partial y}], \qquad (6.7)$$

which establishes the asymptotic validity of using

$$\tilde{E}_{k+1} = h\gamma^*_{k+1}\nabla^{k+1} f_{n+1} . \qquad (6.8)$$

This last argument is not successful for the P_kEC_kE algorithm, where γ_{k-1} is used in (6.1); in our view this clinches the argument in favour of $P_k EC_{k+1} E$.

Having available $\tilde{E}_{k-1}$, E_k, $\tilde{E}_{k+1}$ the decisions about order must now be taken. A simple rule would be to choose the order for which the estimated error is least. Equally simply one could <u>estimate</u> the maximum stepsize that could be used at each order, consistent with satisfaction of the error test. The local error estimates are of orders k, k+1, and k+2; hence the estimates,

$$h.\left(\frac{TOL}{|\tilde{E}_{k-1}|}\right)^{\frac{1}{k}}, \quad h.\left(\frac{TOL}{|E_k|}\right)^{\frac{1}{k+1}}, \quad h.\left(\frac{TOL}{|\tilde{E}_{k+1}|}\right)^{\frac{1}{k+2}} \tag{6.9}$$

can be formed for the maximum stepsizes. Having chosen the order a slightly smaller stepsize is actually attempted, to avoid having many rejected steps. The above discussion shows that a decision to reduce the order can be taken after a rejected step since it can be based on the predicted differences, (6.5).

In practice the above strategies are modified in important respects, based on heuristic arguments. There is a danger in allowing the order to become too high, since errors in the values f_r cause sufficiently high differences to lose any mathematical significance. If the order is mistakenly allowed to become too high then the differences involved in the decision process are meaningless. Also important is the fact that the regions of absolute stability decrease rapidly with increasing order and the lower order methods involve less computation. Consequently it is advisable to weight all the decisions in favour of reducing the order, so that this always occurs when the decisions are close. What usually happens if the order is increased to the point where the higher differences have lost significance is that several rejected steps, with decreasing stepsizes, are attempted at the same point. The program should check for this and restart the integration using only y_n and f_n. A discontinuity in a particular derivative of the solution can produce erratic behaviour in the differences and the order should be reduced to a point where the differences exhibit smooth behaviour. This is a difficult area of implementation. Nevertheless a sound order selection scheme can be based on the principal arguments above, with some safeguards built in.

For a variable step formulation, using divided differences, or modified divided differences, the same strategy can be used. However, the predicted maximum stepsizes (6.9), are not valid unless the stepsize was constant over the last k steps. In practice, (6.9), is used in some programs. The justification for this is,

(i) it is much too costly to produce significantly better estimates.

(ii) in practice the stepsize is not allowed to exhibit large variations and is often held constant over several steps to economise in the computation of the coefficients g_{ij}; a good strategy is to test for an <u>increase</u> in order only when k equal steps at order k have been completed.

(iii) by attempting a fractionally smaller stepsize than is predicted by (6.9) the method can be made to produce a good step selection scheme.

Given a variable order formulation there is in principle no difficulty in starting the integration. This is done by using (6.1) with $k=1$, and requires a much smaller stepsize than will be needed for most of the computation. The order will be built up to an optimum value with gradual increases in the stepsize to its largest possible value. It is necessary to be less restrictive about increasing the order and stepsize for the first few steps. This would improve efficiency, especially on a difficult problem with several discontinuities in the range, perhaps requiring several restarts.

The point of a variable order/variable step method is not so much that these quantities can be <u>varied</u>, but that they can, automatically, be correctly <u>selected</u>. For much of the integration there should be little fluctuation in either. The program has the flexibility to change if the nature of the solution changes.

7. THEORY OF VARIABLE STEP/VARIABLE ORDER PROGRAMS

We have considered two distinct types of algorithm for using Adams methods. One uses the fixed stepsize form of the method, with stepsize changes effected by interpolation; this is referred to as an interpolation mode. The alternative, divided difference, form is simply referred to as a variable-step mode.

The effect of changing stepsize with the interpolation scheme is analysed by Krogh (1973a). He demonstrates that halving the stepsize at high order can cause the error estimate to increase! If this happens when a step is rejected, because the error estimate is too big, then a reduced step will also be rejected. The situation can quickly get out of hand, several steps may be rejected at the same point and deterioration of the differences necessitate a restart. A non rigorous explanation of how this can happen is given here.

Suppose the values $f_n, f_{n-1}, \ldots, f_{n-k+1}$, on which the interpolation polynomial $P_{k,n}(x)$ is based, are exact values of a <u>smooth</u> function $f(x) \equiv f(x, y(x))$. Suppose that on the (n+1)st step, (6.1), the estimated error E_k is too large. Then, as in §4 the stepsize is reduced to h' and the interpolated values

$$\tilde{f}_{n-i} = P_{k,n}(x_n - ih'), \; i = 0,1,\ldots, k-1$$

are used to form new differences. In our discussion, $h' = h/2$, so the process may be illustrated by the following diagram, where $k=7$.

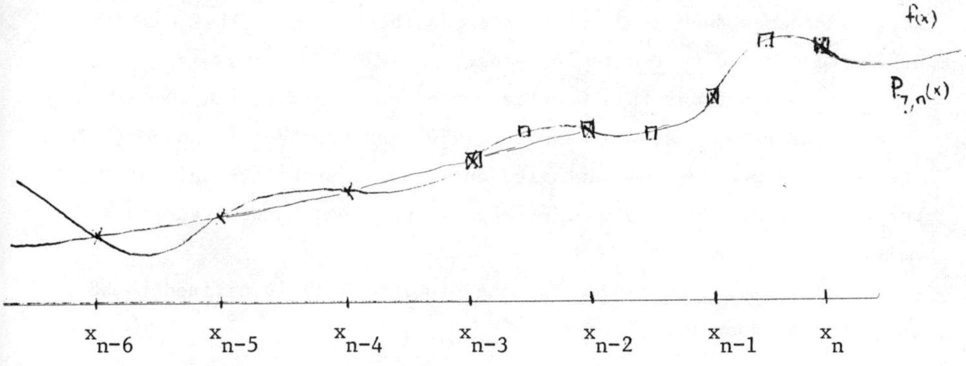

x = back values used before halving, ▫ = back values used after halving.

The values $\tilde{f}_n, \tilde{f}_{n-2}, \tilde{f}_{n-4}$ are exact values of $f(x)$ but the values $\tilde{f}_{n-1}, \tilde{f}_{n-3}, \tilde{f}_{n-5}$ are in error due to interpolation. Using the error term in interpolation theory, the largest error is expected in $\tilde{f}_{n-1}$. We have, in this case,

$$f(x_n - \tfrac{1}{2}h) = \tilde{f}_{n-1} - \delta \,.\, h^7 \; f^{(7)}(\zeta)$$

where $\delta = 33.2^{-11} \simeq 2^{-6}$. Approximation of the derivative by a backward difference leads to

$$f(x_n - \tfrac{1}{2}h) \simeq \tilde{f}_{n-1} - 2^{-6}\nabla^7 f_n$$

When the step is attempted at the halved stepsize the expected value of the seventh backward difference, used in the error estimate, (6.1), assuming no interpolation error, is $(\tfrac{1}{2})^7\nabla_p^7 f_{n+1}$; 2^{-7} times the value obtained for this difference when the step of size h was rejected. The above error in $\tilde{f}_{n-1}$ will alone cause a perturbation of magnitude $21 \times 2^{-6}\nabla^7 f_n$ to the expected value. In this case the estimated error would probably decrease; but by a factor of about $\frac{1}{6}$ rather than the expected $\frac{1}{256}$. (If this reduction is not sufficient to satisfy the error test then a further reduction tends to make things worse). For larger values of k the estimated error is likely to increase when the stepsize is halved. This is a weakness of the interpolation modes. Krogh (1973a) develops modifications, used on the next two steps after halving, to smooth the differences. It should be clear that sufficiently small stepsize changes cannot cause this type of instability and that large increases such as doubling are possible since in this case the differences are expected to increase.

The remainder of this Section represents a simplified summary of some of the results of Gear and Tu (1974) and Gear and Watanabe (1974). These papers present some detailed analysis of stability and convergence for variable step/variable order algorithms based on general linear multistep methods.

A step selection scheme for the problem (1.1) is defined to be a function θ such that

$$h_{n+1} = h\theta(x_n, h), \; h_{n+1} = x_{n+1} - x_n \;,$$

where, for all $h > 0, \; a \leqslant x \leqslant b$,

$$1 \geqslant \theta(x, h) > 0.$$

As h is reduced to zero the maximum stepsize is reduced to zero.

An order selection scheme, in the context of the present discussion, is simply a rule, R, for deciding which of the available orders is to be chosen for the next step.

A method is convergent with respect to θ and R if the computed solution y_n converges to $y(x_n)$ for any $a < x_n < b$ as $h \to 0$.

The classical studies of convergence deal with the case $\theta(x, h) \equiv 1$ and fixed order, Chapter 2. The results below apply equally to any fixed predictor-corrector algorithms and the algorithm of iteration to convergence, using the Adams methods.

Theorem 1

The variable step mode is convergent with respect to any step selection scheme θ and any order selection scheme R.

Theorem 2

The interpolation mode is convergent with repect to any step selection scheme θ and order selection scheme R with the following properties:- there exists a finite constant M such that, in any M consecutive steps there are k steps of constant stepsize using a fixed k-step algorithm, for some value k.

It appears that extra conditions are needed to prove convergence for the interpolation mode. The practical implication of Theorem 2 is that an implementation should hold the order and stepsize constant for k steps after a change. However, this cannot always be achieved since a failure in the error test may necessitate a reduction in the stepsize, at any point.

Consideration of commonly used step selection schemes leads to a different kind of result. A fixed order Adams method, in the interpolation mode, is convergent with respect to a step-selection scheme θ which produces small changes, in the sense that

$$\frac{h_{n+1}}{h_n} = 1 + O(h) \quad \text{as} \quad h \to 0 \tag{7.1}$$

The usual techniques for stepsize selection are based on keeping the local error (or local error per unit step) constant and satisfy (7.1) almost everywhere. To prove convergence at variable order, it still appears necessary to keep the order constant for k steps after a change.

The theory and practice of variable step/order Adams methods are not yet fully developed. The variable step mode, using divided differences or modified divided differences, appears to be the best choice at present. The main argument against it is the non trivial cost of computing the variable coefficients on each step. This is not a strong argument for systems of equations. Neither does one have the difficulties, practical and theoretical, concerning step and order changes for the interpolation mode that have been discussed in this Section. One can devise an excellent program using the interpolation mode but the implementation needs to be more complicated, and conceptually less clear, than that for the variable step mode.

7

EXTRAPOLATION METHODS

1. INTRODUCTION

Extrapolation has already been mentioned (chapter 3) in connection with both local and global error estimates. In such estimates only one extrapolation step was used. For example, if we are given two approximations of the form,

$$I_1 \equiv I_0 + E_1(h)$$

and

$$I_2 \equiv I_0 + E_2(h),$$

then we assume the errors E_1 and E_2 have a particular form say:

$$E_1(h) = a_1 h + O(h^2)$$

and

$$E_2(h) = a_2 h + O(h^2).$$

Using these two error expansions it is possible to derive an improved approximation to I_0 or at least to estimate $E_1(h)$.

The methods given in this chapter will be based on repeated use of this form of extrapolation at each stage of the calculation. Such repeated extrapolation is called Richardson extrapolation or deferred approach to the limit. To emphasise that extrapolation is used at each step it is often called local extrapolation.

2. RICHARDSON EXTRAPOLATION

The standard formulation of the Richardson extrapolation procedure is to eliminate the leading terms in the series expansion

$$F(h) \equiv F_0 + \alpha_1 h + \alpha_2 h^2 + \alpha_3 h^3 + \dots \qquad (2.1)$$

which is evaluated for $h=h_i>0$ $(i=0,..,m)$; where h is a small parameter - step length - which governs the numerical computation. The algorithm used, often called Neville's algorithm, is to set up a tableau.

Tableau 1

$$\begin{array}{llll}
P_0^{[0]} & & & \\
 & P_0^{[1]} & & \\
P_1^{[0]} & & P_0^{[2]} \cdots & \\
 & P_1^{[1]} & & P_0^{[m]} \\
P_2^{[0]} & \vdots & \cdots & \\
\vdots & P_{m-1}^{[1]} \cdots & & \\
P_m^{[0]} & & &
\end{array}$$

The entries of the tableau are determined column by column such that

$$P_i^{[0]} \equiv F(h_i) \qquad (i=0,\dots,m)$$

and

$$P_i^{[j]} = P_{i+1}^{[j-1]} + \frac{P_{i+1}^{[j-1]} - P_i^{[j-1]}}{\left\{\dfrac{h_i}{h_{i+j}}\right\} - 1}, \quad \left\{\begin{matrix} j=1,\dots,m \\ i=0,\dots,m-j \end{matrix}\right\}. \qquad (2.2)$$

If we define $P_i^{[j]}(h)$ by replacing h_i and h_{i+j} in (2.2) by h_i-h and $h_{i+j}-h$, it is obvious that as $P_i^{[0]}(h)$ is a constant $P_i^{[j]}(h)$ is a polynomial of degree j such that

$$P_i^{[j]}(h_i) = P_i^{[j-1]}(h_i) = \dots = P_i^{[0]}(h_i) = F(h_i),$$

that is $P_i^{[j]}(h)$ is an interpolating polynomial. So that $P_i^{[j]}$ is simply the value of the interpolating polynomial defined by $F(h_i),\ldots,F(h_{i+j})$, at $h=0$. It follows directly that the error in the extrapolation algorithm (2.2) is given by

$$F_0 - P_i^{[j]} = h_i h_{i+1} \ldots h_{i+j} E_i^{[j]}. \tag{2.3}$$

The coefficient $E_i^{[j]}$ in the remainder term can be expressed in many ways as a divided difference (Davis 1963) or as

$$\frac{F^{(j+1)}(\xi_i^{(j)})}{(j+1)!} \quad \text{where}$$

$\xi_i^{(j)}$ is contained in $(0, \max(h_i,\ldots,h_{i+j}))$. Usually the sequence $\{h_i\}$ is of the form

$$\{h_0, \frac{h_0}{2}, \frac{h_0}{4}, \frac{h_0}{8}, \frac{h_0}{16}, \ldots\}, \tag{2.4}$$

$$\{h_0, \frac{h_0}{2}, \frac{h_0}{3}, \frac{h_0}{4}, \frac{h_0}{5}, \ldots\} \tag{2.5}$$

or

$$\{h_0, \frac{h_0}{2}, \frac{h_0}{3}, \frac{h_0}{4}, \frac{h_0}{6}, \frac{h_0}{8}, \frac{h_0}{12}, \ldots\}, \tag{2.6}$$

thus the remainder term in (2.3) is $O(h_0^{j+1})$.

The sequence (2.4) leads to accurate results but is often expensive on a computer as halving the step size invariably doubles the work. The sequence (2.5) is cheaper to compute but leads to an unstable form of the extrapolation algorithm as the denominator

$$\frac{h_i}{h_{i+1}} - 1$$

tends to zero. The sequence (2.6) proposed by Bulirsch and Stoer (1964) has the advantages of the other two since it leads to a stable algorithm but does not double the cost of the calculation for each additional row of the Tableau.

It should be emphasised that the success of the extrapolation algorithm (2.2) depends on the existence of the series expansion (2.1). If it is known that the expansion is of the form

$$F(h) = F_0 + \beta_1 h^{\gamma} + \beta_2 h^{2\gamma} + \beta_3 h^{3\gamma} + \ldots, \tag{2.7}$$

the algorithm can be modified to take advantage of this. That is we use polynomials in h^{γ} (usually $\gamma=2$, occasionally $\gamma=4$) and (2.2) becomes

$$P_i^{[0]} = F(h_i) \qquad (i=0,\ldots,m)$$

and

$$P_i^{[j]} = P_{i+1}^{[j-1]} + \frac{P_{i+1}^{[j-1]} - P_i^{[j-1]}}{\left\{\dfrac{h_i}{h_{i+j}}\right\}^{\gamma} - 1} \qquad \begin{matrix}(j=1,\ldots,m)\\(i=0,\ldots,m-j)\end{matrix} \tag{2.8}$$

3. RATIONAL EXTRAPOLATION

If $\gamma=2$, (2.8) is equivalent to approximation by polynomials of successively higher degree in h^2. In many circumstances in numerical analysis it is found that if F is given by (2.8), more accurate approximations can be obtained if the interpolation is in terms of rational functions (Rice 1964) rather than polynomials. Bulirsch and Stoer (1964) developed an extrapolation algorithm based on rational interpolation.

Tableau 2

	$R_0^{[0]}$			
$R_1^{[-1]}$		$R_0^{[1]}$		
	$R_1^{[0]}$		$R_0^{[2]}$	
$R_2^{[-1]}$		$R_1^{[1]}$		
	$R_2^{[0]}$			
$\vdots$	$\vdots$	$\vdots$		$R_1^{[m]}$
	$R_{m-1}^{[0]}$			
$R_m^{[-1]}$		$R_{m-1}^{[1]}$		
	$R_m^{[0]}$			

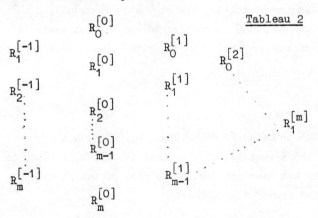

The entries in the rational extrapolation tableau are evaluated column by column as (c.f.(2.8))

$$R_i^{[-1]} = 0, \quad (i = 1,\ldots,m)$$
$$R_i^{[0]} = F(h_i), \quad (i = 0,\ldots,m) \tag{3.1}$$

and

$$R_i^{[j]} = R_{i+1}^{[j-1]} + \frac{R_{i+1}^{[j-1]} - R_i^{[j-1]}}{\left\{\frac{h_i}{h_{i+j}}\right\}^2 \left[1 - \frac{R_{i+1}^{[j-1]} - R_i^{[j-1]}}{R_{i+1}^{[j-1]} - R_{i+1}^{[j-2]}}\right] - 1}.$$

$$(j = 1,\ldots,m)$$
$$(i = 0,\ldots,m-j)$$

It can be shown (Stoer 1961) that $R_i^{[j]}$ is equivalent to interpolating $F(h)$ by a function of the form

$$R_i^{[j]}(h) = \begin{cases} \dfrac{a_0+a_2h^2+\ldots+a_jh^j}{b_0+b_2h^2+\ldots+b_jh^j} & (j \text{ even}) \\[2ex] \dfrac{a_0+a_2h^2+\ldots+a_{j-1}h^{j-1}}{b_0+b_2h^2+\ldots+b_{j+1}h^{j+1}} & (j \text{ odd}) \end{cases}$$

(i.e. $\dfrac{a_0}{b_0}, \quad \dfrac{a_0}{b_0+b_2h^2}, \quad \dfrac{a_0+a_2h^2}{b_0+b_2h^2}, \ldots$).

Another way of computing the $R_i^{[j]}$ has been given by Wuytack (1971), it uses continued fractions and involves fewer arithmetic operations, but it only computes alternate terms, viz. $i+j$ even.

4. THE GRAGG-BULIRSCH-STOER METHOD

In order to apply local extrapolation to the solution of initial value problems it is necessary to use as a basis for the extrapolation, a numerical method for which an asymptotic expansion of the form (2.7) is valid. That is if $y(x)$ is the solution of the differential equation subject to $y(x_0)=y_0$ and $y_n (n=0,1,\ldots,N)$ is the approximation solution at $x_n=x_0+nh$, then for any fixed point $X = x_0+Nh$ (Nh=constant) the approximate solution $y(X;h)$ has an expansion of the form (say)

$$y(X;h) = y(X) + \beta_1 h^2 + \beta_2 h^4 + \beta_3 h^6 + \ldots \quad . \tag{4.1}$$

The existence of such expansions was used, but not proved in chapters 3 and 5 when the local truncation error was estimated by comparing the numerical results after one step of size h, for which (4.1) is valid, with the corresponding results after two steps with step size $\frac{h}{2}$, for which the expansion is

$$y(X;\tfrac{h}{2}) = y(X) + \beta_1 (\tfrac{h}{2})^2 + \beta_2 (\tfrac{h}{2})^4 + \ldots \quad .$$

The existence of asymptotic expansions of the form (4.1) has been investigated by Gragg (1965) and he showed that, for example, it is not possible to base an extrapolation method on the trapezoidal rule since the asymptotic expansion (4.1) is only valid if the difference equations are solved <u>exactly</u>. This is clearly not possible in general as it would require an infinite number of corrections. Gragg has shown that asymptotic expansions exist if the method used is symmetric;

$$\rho(z) + z^k \rho(z^{-1}) = \sigma(z) - z^k \sigma(z^{-1}) = 0,$$

provided the starting values are chosen correctly. Extrapolation cannot be based on implicit symmetric methods for the reason given above for the trapezoidal rule and so the most popular extrapolation algorithm (Bulirsch and Stoer (1966)) is based on the simplest explicit symmetric multistep formula, the <u>midpoint rule</u>:

$$y_{n+2} = y_n + 2hf(x_{n+1}, y_{n+1}). \tag{4.2}$$

Gragg has shown that symmetry is preserved and the expansion (4.1) is valid if the approximation to y(x) is taken to be y(X;h) where

$$y(X;h) = y_N$$

or $$y(X;h) = y_{N-1} + hf(X,y_n)$$

or (usually)

$$y(X;h) = \frac{1}{2}\{y_N+y_{N-1}+hf(X,y_N)\}, \qquad (4.3)$$

provided the additional starting value is given by

$$y_1 = y_0+hf(x_0,y_0). \qquad (4.4)$$

It is this modified midpoint rule that is used in the extrapolation method now called the Gragg-Bulirsch-Stoer (G.-B.-S) method. Computer procedures based on the G.-B.-S. method have been published in ALGOL 60 by Bulirsch and Stoer (1966) and in FORTRAN by Fox (1971) and Gear (1971b). It has been shown by Stoer (1974) that these can be significantly improved.

Gragg (1965) observed that if the extrapolation algorithm (2.8) (or (2.2)) with $F(h_i) = y(X;h_i)$ given by (4.2) - (4.4) is equivalent to a Runge-Kutta method if the basic steplength Nh and the order of the extrapolation m are both fixed. It follows therefore that the stability of the numerical method under such circumstances is guaranteed. Stetter has shown (Lambert (1973)) that polynomial extrapolation has weak stability properties that compare favourably with other methods.

5. IMPLEMENTATION OF RECURRENCE RELATIONS

The extrapolation formula (3.1) is equivalent to

$$R_i^{[j]} = R_{i+1}^{[j-1]} + D_i^{[j]} \qquad (5.1)$$

and by rearranging the differences it follows that

$$D_i^{[j]} = \frac{C_{i+1}^{[j-1]} \quad W_{i+1}^{[j-1]}}{\left\{\dfrac{h_i}{h_{i+j}}\right\}^2 D_i^{[j-1]} - C_{i+1}^{[j-1]}} \qquad (5.2)$$

where

$$C_i^{[j]} = R_i^{[j]} - R_i^{[j-1]}$$

and

$$W_i^{[j]} = R_i^{[j]} - R_{i-1}^{[j]}$$

i.e.
$$W_i^{[j]} = C_i^{[j]} - D_{i-1}^{[j]} . \tag{5.3}$$

The extrapolation formula (3.1) can be written as

$$R_i^{[j]} = R_i^{[j-1]} + \frac{\left\{\dfrac{h_i}{h_{i+j}}\right\}^2 \left[1 - \dfrac{R_{i+1}^{[j-1]} - R_i^{[j-1]}}{R_{i+1}^{[j-1]} - R_{i+1}^{[j-2]}}\right] (R_{i+1}^{[j-1]} - R_i^{[j-1]})}{\left\{\dfrac{h_i}{h_{i+j}}\right\}^2 \left[1 - \dfrac{R_{i+1}^{[j-1]} - R_i^{[j-1]}}{R_{i+1}^{[j-1]} - R_{i+1}^{[j-2]}}\right] - 1}$$

and so $C_i^{[j]}$ is given by the recurrence relation

$$C_i^{[j]} = \frac{\left\{\dfrac{h_i}{h_{i+j}}\right\}^2 D_i^{[j-1]} \quad W_{i+1}^{[j-1]}}{\left\{\dfrac{h_i}{h_{i+j}}\right\}^2 D_i^{[j-1]} - C_{i+1}^{[j-1]}} . \tag{5.4}$$

Thus the algorithm is defined as

$$C_i^{[0]} = D_i^{[0]} = y(X;h_i) \qquad (i=0,\ldots,m)$$

and

$$W_i^{[0]} = y(X;h_i) - y(X;h_{i-1}), \qquad (i=1,\ldots,m)$$

then $D_i^{[j]}$ given by (5.2), $C_i^{[j]}$ given by (5.4) and $W_i^{[j]}$ given by (5.3) for $(j=1,\ldots,m;\ i=0,\ldots,m-j)$. The ordering of the calculation is indicated in tableau 3.

Tableau 3

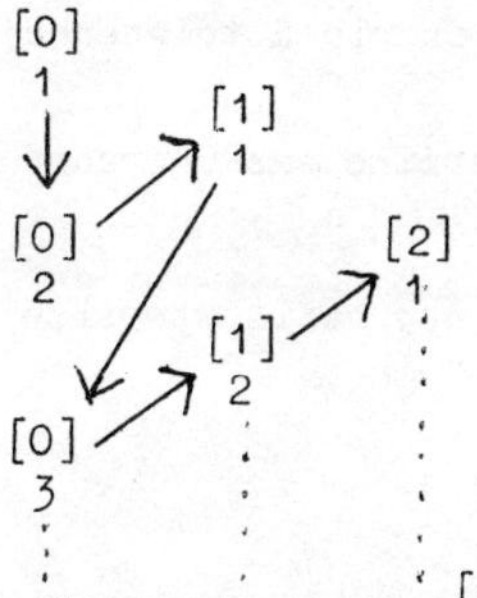

The intermediate values of $C_k^{[i-k]}$ do not have to be stored as $D_k^{[i-k]}$ and $C_k^{[i-k]}$ depend on $C_{k+1}^{[i-1-k]}$ $(k=1,\ldots,i-1)$ and no others, it is however necessarily to store a single array containing $D_k^{[i-k]}$ in order to compute $D_k^{[i+1-k]}$ and $W_k^{[i+1-k]}$ $(k=1,\ldots,i+1)$. The extrapolated values $R_i^{[j]}$ are then determined from

$$R_k^{[i-k]} = \sum_{l=k}^{i} D_l^{[i-l]}$$

as required.

6. IMPLEMENTATION OF STEP CONTROL MECHANISM

Extrapolation can be thought of as a method with variable order (columns of the tableau), variable step size (that is the basic step size h_0) and variable number of stages (rows of the tableau) practical implementation must therefore involve a choice between increasing the order or the number of stages or decreasing the step size to achieve the desired accuracy. Since the sensitivity to round off errors increases with the order of extrapolating, the first implementation (Bulirsch and Stoer 1966) which was developed on a machine with a 40 bit mantissa, limited the order to 6, ($R_i^{[j]}$; $j\leqslant 6$). Later procedures (Fox 1971, Gear 1971b) also use this limit as a standard.

The original step control mechanism was rather crude but it was found by Hull et al. (1972) to perform very well in practice. More will be said about the comparison in the next chapter. If the desired accuracy was achieved without going to 6th order extrapolation; that is for some $i<6$ and $1\leqslant k\leqslant i$,

$$-\frac{1}{s}\,|R_k^{[i-k]} - R_k^{[i-k-1]}| < \epsilon, \text{ (prescribed tolerance)} \qquad (6.1)$$

where $s \approx \max_{x\in[x_0,X]} |y(x)|$ then the basic stepsize was increased.

The recommended rule of thumb was that the new basic stepsize h'_0 was given by

$$h'_0 = 1.5\,h_0. \qquad (6.2)$$

Alternatively if the desired accuracy had not been achieved after 9 stages, that is

$$\frac{1}{s}\,|R_4^{[6]} - R_3^{[6]}| > \epsilon$$

then the basic step size was halved and the extrapolation repeated. If the desired accuracy was achieved after m stages ($6 < m \leq 9$) then the basic step size recommend for the next step was modified. As Gragg (1965) showed that the error in $R_{m-6}^{[6]}$ is $O(h_{m-6}^2 \ldots h_m^2)$ (c.f. (2.3)), if the new basic stepsize is chosen so that the next step $R_1^{[6]}$ will be sufficiently accurate then h'_1 is chosen such that

$$(h'_0 \ldots h'_6)^2 \approx (h_{m-6} \ldots h_m)^2.$$

If the sequence $\{h_i\}$ is given by (2.6) then

$$\frac{h_{m+1}}{h_m} \approx 0.6$$

and so the new basic stepsize is taken such that

$$h'_0 = 0.9(0.6)^{m-6}h_0. \qquad (6.3)$$

A more recent version (Stoer 1974) leads to a significant improvement particularly in those problems where large changes in the step size are called for but the code has not appeared in print yet, it is said to be intended for publication in Numer. Math. The significant modifications are:

a) The maximum order is not fixed at 6, it is

$$l = \left[\frac{p+2.5}{2} \right]$$

where [ ] indicates 'integer part' and p is the number of significant figures required ($\epsilon=10^{-p}$).

b) If the desired accuracy is achieved without going to l-th order extrapolation; that is if

$$\frac{1}{s} \left| R_0^{[j]} - R_0^{[j-1]} \right| \leq \epsilon(n_j^2 - 2) \quad (j \leq l)$$

where $n_j = \dfrac{h_0}{h_j}$, then the recommended basic step size is increased. Whereas if the desired accuracy is reached after $j(>l)$ stages; that is if

$$\frac{1}{s} \left| R_{j-l}^{[l]} - R_{j-1-l}^{[l]} \right| \leq \epsilon\left(\frac{n_j}{n_{j-l-1}} \right)^2, \quad (j>l)$$

then the recommended basic step size is decreased. The modification formulae are rather complicated but they are given by Stoer (1974).

c) If convergence is not sufficiently rapid the extrapolation is restarted with a smaller value of h_0 (details in reference).

8

TESTING PROGRAMS AND COMPARISON OF METHODS

1. INTRODUCTION

The number of factors that can be taken into account when evaluating programs is very large and their relative importance is not the same for all users. Krogh (1972) lists 87 items that could be considered in an evaluation of programs for integrating ordinary differential equations. The main categories are indicated here.

It is obvious that for any program reliability is of major importance. A program is reliable if it obtains the solution to the requested accuracy, whenever this is possible. This must be 'proved' in two ways (i) thorough testing on a comprehensive set of test problems (ii) theoretical results concerning the basic approximation processes and also on the algorithms as implemented. Some modern programs have complex strategies and the theory of implementation is difficult, yet some progress has been made. The program should be robust in the sense that it should only provide a solution when the problem is properly posed. If, too much accuracy is requested for the word-length of a particular machine, the program should be capable of recognizing this and returning an appropriate message.

Equally important is a reasonable degree of efficiency. Fairly extensive testing is required to enable one to draw up guidelines stating which program is expected to solve which types of problems in least time.

It is necessary to make programs easy to use since human effort should not be neglected in the overall cost of solving a problem. It is arguable important to provide some programs that allow the experienced user a wide range of options such as allowing him to modify the program or control the choice of stepsize or method employed by the program. The documentation provided with the program will have a significant effect on its usefulness.

It is not possible to take account of all the factors that could be considered in a comparison, in a quantitative way. The main features that can be so assessed are reliability and efficiency, for a particular class of problems.

2. TESTING A PROGRAM

Krogh (1973b) gives details of a scheme for testing a program. The philosophy behind his approach is to stretch the program under consideration to various extremes, to examine the performance of the method under such conditions as well as in intermediate situations. We outline the main features of his scheme below.

He gives a set of fourteen differential equation problems, each of which is selected for a specified reason. They include examples where for part of the range local accuracy limits the stepsize and and where stability is the limiting factor for the remainder of the integration. One problem is included where frequent changes in the stepsize are necessary and another where a constant stepsize should be used. Some require control of absolute error and others relative error.

Each problem is required to be solved with local accuracy requirements 10^s, for $s = 1,0,-1,\ldots,-20$, on a machine with about 18 decimal digits. No external limits are imposed on the stepsize since part of the testing is to see how the method copes with step selection, (some programs require that the user specify a maximum stepsize, but in many cases it is very difficult for users to make this choice correctly.) The output of the solution is required at three points in the range.

The statistics provided for each problem at each accuracy requirement, are the error at each output point (absolute or relative as appropriate) and the number of derivative evaluations. Some important questions that can be answered from the statistics are:

(i) is the global error consistently less than the user's error tolerance?

(ii) does the global error decrease smoothly with the users error tolerance? This is important for the estimation of global error in practice, which is often done by integrating the equations at different tolerances.

(iii) does the program return correct diagnostics when too much accuracy is requested?

(iv) is the program stable when very little accuracy is requested?

The number of derivative evaluations is a partial indication of efficiency, particularly relevant for problems where such evaluations are expensive to compute. The test is severe, which it must be to make up for the lack of an adequate theory of modern programs.

Similar testing of several methods has been carried out by other practitioners, for example in Manchester in connection with the development of programs for the NAG library (Walsh 1974). If the answers to the above questions are in the affirmative then one has reasonable confidence in the reliability and robustness of the program. Krogh's idea was to try to set a standard, regarding a set of problems and presentation of results, for this type of testing and as such it is extremely useful.

3. COMPARISON OF METHODS AND PROGRAMS

A quite different approach to testing was used in the comparisons presented by Hull, Enright, Fellen and Sedgwick (1972). As implementations improved these comparisons were updated (Enright, Bedet, Farkas and Hull 1974). They were made fully automatic by designing a program DETEST to carry out the tests.

DETEST is described in Hall, Enright, Hull and Sedgwick (1973) and has been made generally available. There are twenty-five differential equation systems, each of which has to be solved at three different tolerances (TOL) 10^{-3}, 10^{-6}, 10^{-9}. For each of the seventy-five problems thus obtained DETEST outputs one line of primary statistics used for assessing reliability and efficiency. These are

(i) Time; total time in seconds to integrate over the range.

(ii) Overhead; the total time, excluding derivative evaluations.

(iii) The number of derivative evaluations.

(iv) The number of integration steps.

(v) The maximum error. On the nth step of the method, $h_n = x_n - x_{n-1}$, the true <u>local error</u>, ln, (Chapter 1) is computed by DETEST. To be tested by DETEST it is necessary for the method to try and keep local error per unit step less than TOL. The maximum error is defined to be

$$\max_n \frac{|ln|}{h_n * TOL} .$$

(vi) The proportion of steps on which $|ln| > h_n * TOL$.

(vii) The proportion of steps on which $|ln| > 5h_n * TOL$.

It will be seen that no attempt is made to examine global error. The approach to the testing of reliability is to say that since methods, in practice, attempt to control local error, they are reliable if they succeed in doing this. Given this approach it is natural to require the programs to control local error per unit step. Even if two methods use a different number of steps then, for sufficiently severe tolerances, with this type of error control, the global errors will be approximately equal for many problems. For example, if the solution of the problem

$$y' = -y^2, \quad y(0) = 1, \quad 0 \leqslant x \leqslant 1$$

is obtained using a Taylor Series Method of order p with

(a) error per unit step control

(b) error per step control

one can show, for sufficiently small values of TOL, that the error is approximated by

$$\text{(a)} \quad \frac{TOL}{3}\left(\frac{1}{(1+x)^2} - (1+x)\right)$$

$$\text{(b)} \quad TOL^{p/p+1}\,\frac{p+1}{2p+1}\left(\frac{1}{(1+x)^2} - \frac{1}{(1+x)^{\frac{1}{p+1}}}\right)$$

Using (a) the error is the same for all values of p; for example, with TOL = 10^{-5}, Euler's method requires about 38000 steps and the global error is $\cdot 60 \times 10^{-5}$ at x = 1. The second order Taylor Series method requires 160 steps with global error $.59 \times 10^{-5}$. These values are consistent with expected value.

DETEST is an excellent program for comparing methods; it gives precise measures of cost with a useful breakdown into derivative evaluations and overhead. It does not avoid the need for the kind of testing of a program discussed in §2. Such testing usually persuades practitioners that error per step control is to be preferred in practice. Some testing by Ian Gladwell in Manchester recently shows clearly that a method which is reliable as tested by DETEST may at some tolerances be quite poor with regard to global accuracy, on the same test problems. This is probably because of the form of error control used.

The main features of the comparisons (Enright et al 1973) for the important types of method, which are supported by other testing (Walsh 1974) are

(1) Variable order Adams. These programs require significantly fewer derivative evaluations than all other methods and should always be used when these evaluations are expensive (more than 25 arithmetical operations per derivative.) They require substantially more overhead than Runge-Kutta methods. The evidence from Enright et al (1974) is that the divided difference form is the most efficient way to use the Adams methods.

(2) Runge-Kutta methods. Among a wide variety of possibilities it has become clear from DETEST that the Fehlberg type method/error estimation is the most promising, Chapter 3. At present such methods are the most efficient in terms of total time unless derivative evaluations are expensive. A fourth order Runge-Kutta method has the advantage of simplicity and is the most efficient method for many routine low accuracy calculations.

(3) Extrapolation, Gragg-Bulirsch-Stoer. It is generally agreed that the implementations of this method are not yet as fully developed as for other methods. Nevertheless it has always done well in comparisons with other methods, especially for high accuracy work.

4. CONCLUSION

It appears likely to this author that the variable order Adams methods will develop into the most commonly used methods. By the correct selection of order they can be efficient over a wide range of tolerances. The user's job is easier since he does not have to select the order. It is easy to obtain the solution at the output points, using interpolation, without restricting the stepsize. A special, efficient starting procedure can be designed. It is easy to generalise the methods to higher order equations. They can be imbedded in a more general program which includes methods for stiff systems by a further generalisation of the interpolation derivation. The order of method and the method itself can easily be different for different equations in a system. The main (and very difficult) problem of implementation is to choose the orders/methods/stepsizes quickly and correctly in an automatic way and then, in the interests of reliability, avoid unnecessary changes.

PART 2

STIFF PROBLEMS

9

INTRODUCTION TO STIFF PROBLEMS

In this first chapter on stiff equations, I shall attempt to define what is meant by "stiffness" and to describe the basic problem of numerical stability posed by stiff systems. Some general features arising in the implementation of stable, implicit Runge-Kutta and linear multistep methods for solving stiff systems of equations will be discussed.

1. STIFF INITIAL-VALUE PROBLEMS; STIFFNESS RATIO

"Stiffness" is a property of a mathematical problem (not of the numerical solution method) and we shall take the problem to be an initial-value problem involving a system of s coupled, generally non-linear, first-order ordinary differential equations

$$y' - f(x,y) = 0 \ . \tag{1.1}$$

Let $y(x)$, $x \in [a,b]$, denote the exact solution to equations (1.1) which satisfies the given initial conditions $y(a) = y_0$.

To determine whether or not this initial-value problem is stiff, we need to know something about the nature of the solutions to equations (1.1) in the neighbourhood of the particular solution $y(x)$. In such a neighbourhood, equations (1.1) may be closely approximated by the linearized, variational equations

$$y' - J(x)\left\{y - y(x)\right\} - f(x,y(x)) = 0 \ , \tag{1.2}$$

where $J(x)$ denotes the Jacobian matrix of partial derivatives $\partial f/\partial y$, evaluated at $(x,y(x))$. If the variation of $J(x)$ in an interval of x is sufficiently small, the localized eigensolutions of equations (1.2) are approximately the exponentials $e^{\lambda_i x}$, where the $\lambda_i = \lambda_i(x)$ are the local eigenvalues (assumed distinct) of the Jacobian matrix J(x).

Thus, the solutions y to equations (1.1) in a neighbourhood of the exact solution y(x) at x are of the form

$$y \approx y(x) + \sum_{i=1}^{s} c_i e^{\lambda_i x} \xi_i , \qquad (1.3)$$

where the c_i are constants and the ξ_i are the eigenvectors of J(x). The eigensolutions $e^{\lambda_i x}$ characterize the local response of the system to small changes or perturbations about y(x). We shall assume that the system is locally stable, so that

$$\mathrm{Re}(\lambda_i) < 0 \ (i = 1,2,\ldots,s) ,$$

and the transient eigensolutions decay with increasing x at rates which are proportional to

$$1/\mathrm{Re}(-\lambda_i)$$

termed the local "time constants" of the system. It is the range in the local values of the "time constants" of a problem that provides a measure of stiffness.

<u>Definition 1</u> (cf Lambert, 1973): The initial-value problem

$$y' - f(x,y) = 0 \quad , \quad y(a) = y_0 \quad , \quad x \in [a,b] \quad ,$$

is said to be <u>stiff</u> in an interval $I \subset [a,b]$ if, for $x \in I$,

(1) $\mathrm{Re}(\lambda_i) < 0 \quad (i = 1,2,\ldots,s)$; and

(2) $S(x) = \max_{i=1,s} \mathrm{Re}(-\lambda_i) / \min_{i=1,s} \mathrm{Re}(-\lambda_i) \gg 0$;

where the λ_i are the eigenvalues of $\partial f/\partial y$ evaluated on the solution y(x) at x.

The ratio S(x) may be termed the (local) "stiffness ratio" of the problem (Lambert, 1973). Problems may be considered to be marginally stiff if $S(x)$ is $O(10)$, while stiffness ratios up to $O(10^6)$ are not uncommon in practical problems arising in such fields as chemical kinetics, process control and electrical circuit theory.

A stiff problem is often referred to in the literature as a problem with "widely differing time constants" or as a system "with a large Lipschitz constant". If the derivatives $\partial f/\partial y$ are continuous and bounded, the Lipschitz constant

$$L = ||\partial f/\partial y|| > \rho(\partial f/\partial y) ,$$

where the spectral radius ρ is defined to be

$$\rho = \max_{i=1,s} |\lambda_i| .$$

2. STABILITY FOR STIFF PROBLEMS

Let us now consider the difficulties that arise in attempting to obtain a numerical approximation to the solution $y(x)$ of a stiff problem. The basic problem is that of numerical stability.

For the absolute stability of numerical solutions to the system of equations (1.1), it is necessary to use a step size h, such that every one of the (complex) values $\bar{h}_i = h\lambda_i$ $(i = 1,2,\ldots,s)$, where the λ_i are the eigenvalues of $J(x)$, lies within the region of absolute stability of the numerical method. For methods with finite absolute stability regions, the step size is thus restricted to the order of magnitude of the smallest time constant of the system, and as the range of integration may well exceed the value of the largest time constant, the number of integration steps required may be comparable to the stiffness ratio of the system.

To overcome this stability limitation on the step size, numerical methods have been sought that possess regions of absolute stability that extend to infinity in the half-plane $\mathrm{Re}(h\lambda) < 0$. Several definitions describing stability properties suitable for stiff systems have been proposed.

Definition 2 (Dahlquist, 1963): A numerical method is said to be A-stable if its region of absolute stability contains the whole of the half-plane $\mathrm{Re}(h\lambda) < 0$.

The A-stability property was originally formulated for linear multistep methods, and it proves to be a very restrictive requirement for such methods. Dahlquist (1963) also proved that no explicit linear multistep method could be A-stable, and that the order of an implicit linear multistep method could not exceed two, the trapezoidal rule being the most accurate such method. In consequence, A-stability has been most widely used in studying implicit one-step methods (Ehle, 1968; Axelsson, 1969; Chipman, 1971a) and various explicit generalizations of Runge-Kutta and linear multistep methods (Treanor, 1966; Norsett, 1969b ; Lambert and Sigurdsson, 1972). A number of less demanding stability properties have proved more useful for multistep methods.

Definition 3 (Widlund, 1967): A numerical method is said to be $A(\alpha)$-stable, $\alpha \in (0,\pi/2)$, if its region of absolute stability contains the infinite wedge $|\arg(-\lambda)| < \alpha$ (Figure 1). A method is said to be A(0)-stable if it is $A(\alpha)$-stable for some (sufficiently small) $\alpha \in (0,\pi/2)$.

Note that in the above definition a given eigenvalue λ either lies inside or outside the wedge, regardless of the positive step size used. Widlund proved that no explicit linear multistep method could be A(0)-stable, and showed that implicit linear multistep methods of orders three and four existed that were $A(\alpha)$-stable for any $\alpha < \pi/2$.

The above definitions are concerned only with stability. Gear (1969; 1971b) defines a more complex property, involving both stability and accuracy of approximations to the exponential eigensolutions.

Definition 4 A method is said to be stiffly stable if it is

(a) absolutely stable in the region R_1 $(\mathrm{Re}(h\lambda) \leq D)$ and

(b) accurate in the region R_2 $(D < \mathrm{Re}(h\lambda) < \alpha,\ |\mathrm{Im}(h\lambda)| < \theta)$; (see Figure 2).

The reasoning behind this definition requires some explanation. If the solution $y(x)$ contains transient eigensolutions $e^{\lambda_i x}$, those eigensolutions with small time constants decay rapidly and quickly become negligible, while the eigensolutions with larger time constants may contribute significantly to the solution. Using a stiffly stable

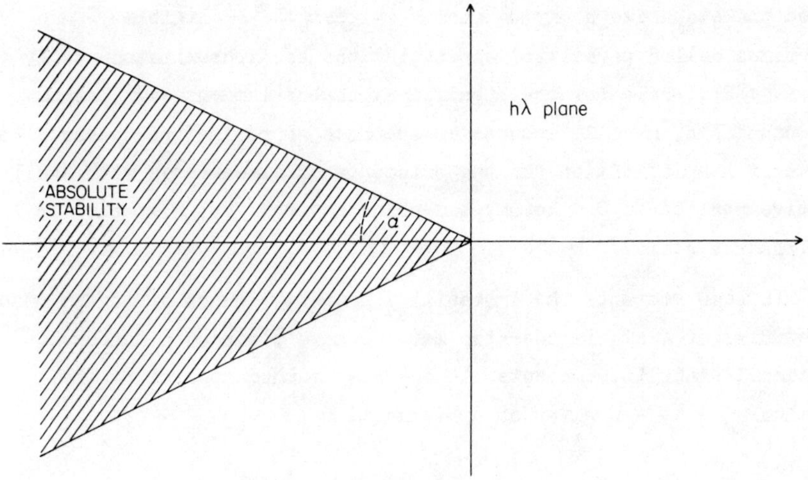

FIG 1—A(α)-stability (Widlund, 1967)

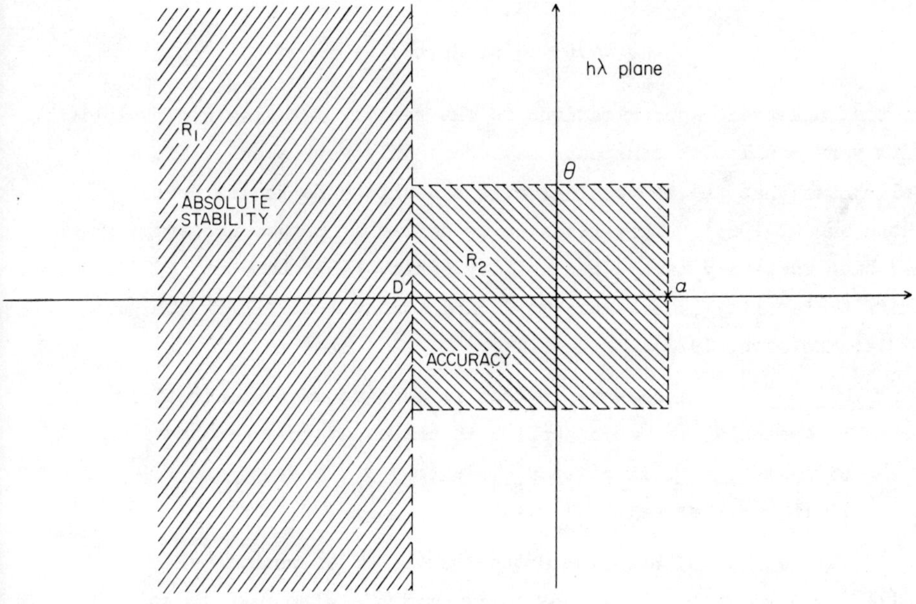

FIG. 2—Stiff stability (Gear, 1971)

method the step size h may be chosen so that the negligible (sometimes called parasitic) eigensolutions are approximated stably with $h\lambda \in R_1$, while for the significant eigensolutions with larger time constants, $h\lambda \in R_2$ ensures an accurate approximation. Gear also allows in his definition for the existence of eigenvalues with small positive real parts $0 < \text{Re}(h\lambda) < \alpha$ which may well occur locally in non-linear systems.

In some respects the A-stability property is not stringent enough, and studies of A-stable one-step methods have produced a number of additional stability concepts. A one-step method applied to the equation $y' - \lambda y = 0$ gives an approximation

$$y_{n+1} = Q(h\lambda)\, y_n$$

to the solution

$$y(x_{n+1}) = e^{h\lambda}\, y(x_n)\ ,$$

and A-stability ensures that $|Q(h\lambda)| < 1$ for all $\text{Re}(h\lambda) < 0$. For many A-stable one-step methods, however, $Q(h\lambda)$ is such that

$$|Q(h\lambda)| \to 1 \text{ as } \text{Re}(h\lambda) \to -\infty\ ,$$

so that numerical approximations to the rapidly decaying eigensolutions with very small time constants may decay only very slowly. It is well-known that the trapezoidal rule has this unfortunate property (Rosenbrock, 1963). This leads us to define a further property, which has been variously termed L-stability (Ehle, 1969; Lambert, 1973), stiff A-stability (Axelsson, 1969), and strong A-stability (Chipman, 1971a; Axelsson, 1972):

Definition 5 A one-step method is said to be L-stable if it is A-stable, and when applied to the equation $y' - \lambda y = 0$ with $\text{Re}(\lambda) < 0$, it gives $y_{n+1} = Q(h\lambda)\, y_n$, where $|Q(h\lambda)| \to 0$ as $\text{Re}(h\lambda) \to -\infty$.

Gourlay (1970) has also noted that one-step methods with $|Q(h\lambda)| \to 1$ as $\text{Re}(h\lambda) \to -\infty$ may prove unstable when used to solve

$$y' - \lambda(x)\, y = 0\ , \qquad (2.1)$$

a test equation appropriate to the variational equations (1.2), and he proposes a modification for the trapezoidal rule which is always stable for equation (2.1).

Another generalization of the test equation $y' - \lambda y = 0$ that is appropriate to non-linear systems is the following scalar form of equations (1.2):

$$y' - \lambda\, y - y(x) \quad - f(x,y(x)) = 0\,. \tag{2.2}$$

Solution of this equation using one-step methods gives a difference equation of the form

$$e_{n+1} = y_{n+1} - y(x_{n+1}) = \alpha(h\lambda).e_n + \beta(h,h\lambda,y(x)). \tag{2.3}$$

Prothero and Robinson (1974) define a stability property, termed S-stability, which generalizes the A-stability concept to equations (2.2), and examine the asymptotic accuracy of one-step methods in the limit $|h\lambda| \to \infty$. Thus the error in the trapezoidal rule in this limit is of the form

$$\beta \approx -\frac{1}{6}h^3 (h\lambda)^{-1}\, y_n''' \,,$$

while for the closely related implicit midpoint rule

$$\beta \approx -\frac{1}{4}h^2\, y_n''\,.$$

One-step methods for which $\beta \to 0$ as $|h\lambda| \to \infty$ are said to be stiffly accurate.

3. IMPLICIT METHODS

In order that a classical linear multistep or one-step method may possess any of the A-stability or stiff stability properties discussed in the previous section, it is necessary for the method to be implicit. An implicit linear multistep method

$$\sum_{j=0}^{k} \alpha_j\, y_{n+j} - h \sum_{j=0}^{k} \beta_j\, f_{n+j} = 0\ ;\ \alpha_k = 1,\ \beta_k \neq 0, \tag{3.1}$$

applied to a non-linear system (1.1) results in a system of s non-linear equations

$$y_{n+k} - h\beta_k f(x_{n+k}, y_{n+k}) - g = 0 , \tag{3.2}$$

with g a known vector, to solve for y_{n+k} at each integration step. The stability properties of the method (3.1) may only be retained by solving equations (3.2) accurately using a convergent iterative method. Simple iteration of the form

$$y_{n+k}^{(m+1)} = h\beta_k f(x_{n+k}, y_{n+k}^{(m)}) - g, \quad m = 0,1,\ldots ,$$

is not practicable, since for convergence we would require

$$h\beta_k L < 1 ,$$

where L is the Lipschitz constant of $f(x,y)$ with respect to y, and for stiff systems this convergence condition imposes just the type of severe restriction on the step size that we are trying to avoid.

The Newton-Raphson method for solving non-linear equations has been found to work successfully (Gear, 1969; Liniger and Willoughby, 1970). Applied to equations (3.2), the method gives at each iteration a set of linear equations

$$\left\{I - h\beta_k J^{(m)}\right\}\left\{y_{n+k}^{(m+1)} - y_{n+k}^{(m)}\right\} = - y_{n+k}^{(m)} + h\beta_k f\,(x_{n+k}, y_{n+k}^{(m)}) + g ,$$
$$m = 0,1,\ldots, \tag{3.3}$$

where $J^{(m)} = \frac{\partial f}{\partial y}(x_{n+k}, y_{n+k}^{(m)})$. Sufficient (but rather complex) conditions for the convergence of the Newton-Raphson method have been formulated by Kantorovich (e.g., see Henrici, 1962). In practice, convergence may usually be obtained without any restriction on the step-size h provided a good initial estimate $y_{n+k}^{(0)}$ can be calculated using a separate predictor.

For large systems of differential equations, the Newton-Raphson method can be very time-consuming, as each iteration requires the re-evaluation of the Jacobian matrix and the solution of the linear system (3.3). To save time, it is common to use a modified Newton-Raphson procedure in which the Jacobian evaluation and an LU decomposition of the matrix $\{I - h\beta_k J^{(0)}\}$ are carried out only for the first iteration ($m = 0$) in an integration step, and this same matrix decomposition is then used in subsequent iterations. Indeed the same iteration matrix may often be used for several integration steps before failure to converge in a limited number of iterations indicates that re-evaluation is necessary. Liniger (1971) has reported the number of convergent iterations needed to achieve a given order of accuracy in the solution of the non-linear equations (3.2).

The accuracy of the predicted values $y_{n+k}^{(0)}$ used to start the Newton-Raphson iteration can have a considerable influence on the number of iterations needed for convergence and on the number of times that the iteration matrix needs to be re-evaluated. Robertson and Williams (1975) have shown that the use of previously-calculated values of the derivatives f in the predictor formula can lead to inaccuracy in the presence of eigensolutions $e^{\lambda x}$ with $\mathrm{Re}(-h\lambda)$ large, and they recommend that predicted values should be obtained by extrapolation through previous values of the solution y only. As the difference between the predicted values $y_{n+k}^{(0)}$ and the solution y_{n+k} are often used in obtaining an estimate of the local truncation error, required for checking accuracy and controlling step size, it is clearly important that the accuracy of the predictors is not impaired by terms that are insignificant in the solution.

Efficient procedures for evaluating the Jacobian and for performing the matrix decomposition or inversion are also important. The partial derivatives $\partial f/\partial y$ are often estimated using a forward-difference approximation

$$\frac{\partial f}{\partial y_j} \approx \frac{f(x,y + \delta_j e_j) - f(x,y)}{\delta_j}, \tag{3.4}$$

where e_j is the normalized jth coordinate vector and δ_j is an increment. This procedure requires (s+1) evaluations of the s derivatives f. Curtis and Reid (1974) have considered the problem of selecting increments δ_j to minimize truncation and round-off errors, but their proposal to use central differences, with forward differences to monitor accuracy, involves (2s+1) evaluations of f. Curtis, Powell and Reid (1974) have pointed out that the constant elements in a Jacobian matrix need not be re-evaluated, while the varying components arising from the non-linearity of f may sometimes constitute a sparsely filled matrix. By using the sparsity pattern, the number of evaluations of f may often be reduced from $(s + 1)$ to $(n_r + 1)$, where n_r is the largest number of non-zero elements in any row of J. Special procedures may also be used for banded Jacobian matrices resulting from the discretization of parabolic partial differential equations.

4. STEP-SIZE STRATEGIES

Using a numerical method that possesses an appropriate A-stability or stiff stability property, the step size h should not be restricted on account of stability or, hopefully, by any convergence requirement in the iterative solution of the implicit equations, so that the choice of step size may be based on the accuracy required in the solution. In solving stiff systems, two step-size strategies are commonly used:

(1) When the effects on the solution of the transient eigensolutions with small time constants are not of interest, an initial step size that is large relative to the small time constants may be used. No attempt is made to approximate the short-term effects accurately, and the stability of the method is relied upon to damp out these transients when calculating the long-range behaviour of the solution. Using this strategy the L-stability property is particularly useful in ensuring quick damping of the most rapidly varying transients.

(2) When an accurate representation of the rapidly varying transients in the solution is required, the initial step size must be comparable with the smallest time constants. As described in the discussion of Gear's stiff stability concept, once the most rapidly varying transients have decayed sufficiently, the step size may be increased to a value comparable with the smallest time constant of a significant eigensolution. Efficient use of this strategy requires the implementation of a variable step-size facility with effective step-size control based on estimates of the local truncation errors.

5. ILLUSTRATIVE EXAMPLE

To illustrate some of the material presented in this lecture, I shall conclude with a brief presentation of a well-known stiff test problem and a description of the solution obtained using a simple A-stable integration method.

The non-linear problem from the field of chemical kinetics, published by Robertson (1966), involves three equations:

$$\begin{aligned} y_1' + 0.04\, y_1 - 10^4\, y_2\, y_3 &= 0\,, \\ y_2' - 0.04\, y_1 + 10^4\, y_2\, y_3 + 3.10^7\, y_2^2 &= 0\,, \\ y_3' - 3.10^7\, y_2^2 &= 0\,, \end{aligned} \tag{5.1}$$

with $a = 0$, $b = 100$ and $y^T(0) = (1,0,0)$. The Jacobian matrix of the system (5.1) has one zero eigenvalue (since $y_1 + y_2 + y_3 = 1$ for all x) and two real negative eigenvalues which vary along the solution. Figure 3 shows the varying "time-constants"

$$\tau(x) = \mathrm{Re}(-1/\lambda(x))\,,$$

and the stiffness ratio $S(x)$, which varies from $O(10^4)$ to $O(10^5)$.

As the eigenvalues are always real, we could use any A(0)-stable method to solve this problem. The solution that I describe is obtained using a variable-step implementation of an A-stable linear one-step method

$$y_{n+1} - y_n - h(0.55\, f_{n+1} + 0.45\, f_n) = 0. \tag{5.2}$$

This method is A-stable giving

$$y_{n+1} = Q(h\lambda).y_n = \left(\frac{1 + 0.45\, h\lambda}{1 - 0.55\, h\lambda}\right).\, y_n$$

when applied to the equation $y' - \lambda y = 0$. Note that $Q(h\lambda) \to -9/11$ as $|h\lambda| \to \infty$. The order of the method (5.2) is one, the principal truncation error being $\frac{1}{20}\, h^2\, y''$.

My implementation of the method (program name STINT) uses a modified Newton-Raphson iteration to solve the non-linear equations, with accurate predicted values used to save on iterations (not more than two iterations per step are needed for the present example) and

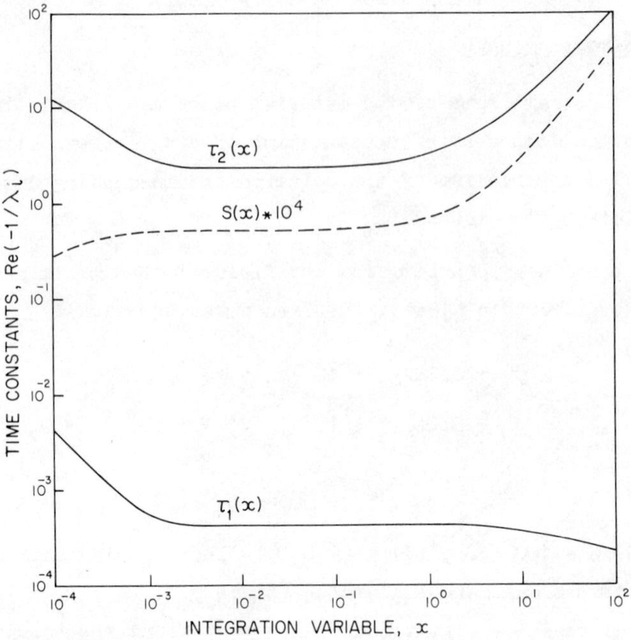

FIG. 3 – Stiffness ratio $S(x)$ and time constants $\tau(x)$ of problem (5.1)

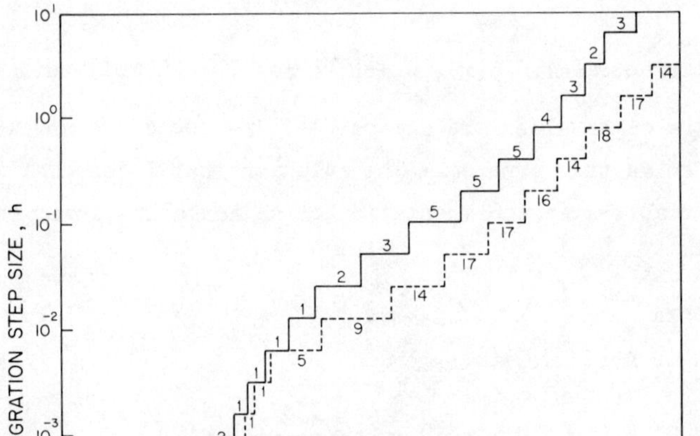

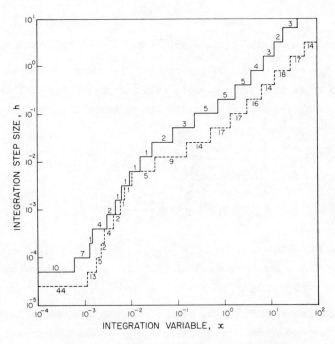

FIG. 4 – Variation of step size and number of steps in solution of problem (5.1).
Solid line, $\varepsilon = 10^{-3}$; broken line, $\varepsilon = 10^{-4}$

to provide accurate local truncation error estimates for selecting the integration step size. The accuracy is controlled using a limit ϵ on the largest relative truncation error per integration step, and the accumulated errors in the solution are estimated and monitored as a safeguard.

Solutions have been obtained for two values of the accuracy parameter, $\epsilon = 10^{-3}$ and 10^{-4}. The initial step size ($h_0 = 10^{-4}$) is chosen to be of comparable order to the smallest time constant. Figure 4 shows the manner in which the step size increases during the integration. Over a range of x of $O(10^{-3})$ the rapidly-varying transient is accurately approximated using a small step size. Once this transient has decayed sufficiently, the step size increases rapidly to a value that gives a sufficiently accurate approximation to the slowly-varying transient, with $|h\lambda|_{max}$ reaching $O(10^4)$.

The amount of computation involved in these runs is summarized in Table 1. For this problem the Jacobian matrix has only been re-evaluated, and the iteration matrix re-inverted, when the step size has been doubled or halved.

Accuracy parameter ϵ	10^{-3}			10^{-4}		
Integration range	0-1	0-10	0-100	0-1	0-10	0-100
No. of successful steps	41	55	65	111	160	214
No. of rejected steps	1	1	2	3	3	3
No. of times step size doubled	12	15	20	12	15	18
No. of Jacobian evaluations and matrix inversions	14	17	23	16	19	22
No. of derivative evaluations	106	141	186	197	304	421

Table 1. Performance of method (5.2) on problem (5.1).

10

IMPLICIT RUNGE-KUTTA AND RELATED METHODS

1. LIMITATIONS OF EXPLICIT RUNGE-KUTTA METHODS

In the search for reliable and efficient one-step methods for stiff problems, explicit Runge-Kutta methods are avoided for two principal reasons. The first is also a consideration for non-stiff problems and is simply that the computational cost, particularly as measured in terms of derivative evaluations increases rapidly as higher order requirements are imposed. The second reason is specific to stiff problems and is concerned with the stability properties of these methods.

As far as order of accuracy is concerned, if we require an order p to be achieved, a set of N_p algebraic conditions must be satisfied, where N_p is the number of rooted trees with no more than p nodes. These conditions, which will be described in detail in Section 2, must be satisfied by the selection of the $s(s+1)/2$ parameters of an explicit s stage method. As it happens, an order p can be achieved with $s=p$ stages for $p=1,2,3,4$ but for higher p the minimum possible value of $s-p$ increases. Furthermore, the derivation of particular methods of high order is exceedingly complicated.

The stability question is concerned with the application of a Runge-Kutta method to the test problem $y'=qy$, where q is a constant (possibly complex) number, using step size h. If $hq=z$, then it is easy to show that $y_n=P(z)y_{n-1}$ where p is a polynomial of degree s (the number of stages). Since $|P(z)| \to \infty$ as $|z| \to \infty$, such a method cannot possibly have an unbounded stability region.

We are thus forced to look at implicit methods if we wish to use Runge-Kutta methods at all for stiff problems. Accordingly, the discussion of the order conditions in Section 2 is in the more general context of implicit methods.

2. ORDER CONDITIONS FOR RUNGE-KUTTA METHODS

We will write the system to be solved as a family of autonomous equations

$$y^{i\prime}(x) = f^i(y^1(x),\ y^2(x),\ldots,y^N(x)) \tag{2.1}$$

$(i=1,2,\ldots,N)$, which we will also write more compactly as

$$y'(x) = f(y(x)) \tag{2.2}$$

Note that there is no loss of generality in restricting ourselves to a system (2.1) or (2.2) in which x does not appear as one of the arguments of $f^1, f^2,\ldots, f^N$ since, if necessary, an additional differential equation can be added to the family for which the solution is x. We will consider the computation of step number n of a Runge-Kutta method with step size h.

Let y_{n-1} denote the solution vector at the beginning of this step, let $Y_1, Y_2,\ldots, Y_s$ denote the intermediate values computed within the step and let y_n denote the solution vector at the end of the step. Then $Y_1, Y_2,\ldots, Y_s$ and y_n are given by

$$\left.\begin{aligned} Y_1 &= y_{n-1} + h\sum_{j=1}^{s} a_{1j}f(Y_j) \\ Y_2 &= y_{n-1} + h\sum_{j=1}^{s} a_{2j}f(Y_j) \\ Y_s &= y_{n-1} + h\sum_{j=1}^{s} a_{sj}f(Y_j) \end{aligned}\right\} \tag{2.3}$$

$$y_n = y_{n-1} + h\sum_{j=1}^{s} b_j\, f(Y_j) \tag{2.4}$$

where $a_{11}, a_{12}, \ldots, a_{ss}, b_1, b_2, \ldots, b_s$ are numerical constants that characterise a method. We will represent a particular method by writing these constants arranged in an array as follows

$$\begin{array}{c|ccc} c_1 & a_{11} & \cdots & a_{1s} \\ c_2 & a_{21} & \cdots & a_{2s} \\ \vdots & \vdots & & \\ c_s & a_{s1} & \cdots & a_{ss} \\ \hline & b_1 & \cdots & b_s \end{array}$$

where for $i=1,2,\ldots,s$ the value of c_i is defined as $\sum_{j=1}^{s} a_{ij}$. We wish to obtain conditions that the vector y_n differs from $y(x_{n-1}+h)$ by $O(h^{p+1})$ as $h \to 0$ where $y(x_{n-1}) = y_{n-1}$.

Let T denote the set of all rooted trees. Thus the following are members of T where the diagrams for these (rooted) trees are drawn by convention with the root at the bottom.

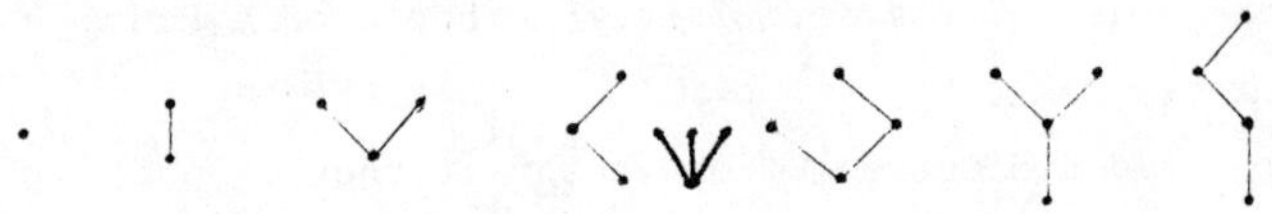

We will need to refer to certain particular trees so we designate by the names t_k, t^k and $t_{k,1}$ the trees with k, k and k+1 nodes respectively with these diagrams:

$t_k =$ 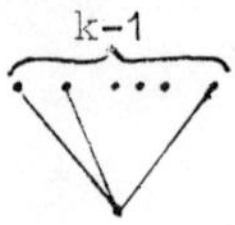

$t^k =$ 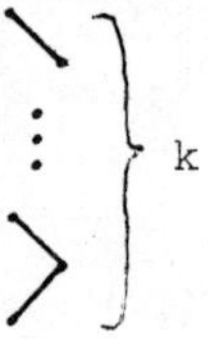

$t_{k,1} =$ 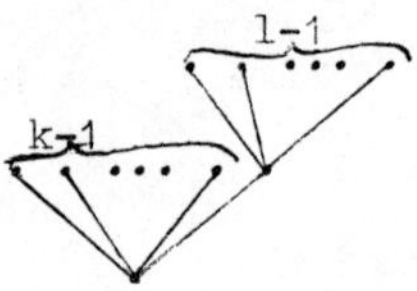

For each $t \in T$ we may form a quantity $F(t)(y_{n-1})$ which is given by associating with each node of t the n-linear operator $f^{(n)}(y_{n-1})$, which we will henceforth write as $f^{(n)}$ for brevity, where n is the number of upward growing branches from this node (in the case of terminal nodes we identify the 0-linear operator $f^{(0)}$ as the constant vector f) and then interpret the tree as an operation diagram. For example,

$$F(t_k)(y_{n-1}) = f^{(k-1)}(\underbrace{f, f, \ldots, f}_{k-1})$$

$$F(t^k)(y_{n-1}) = (f^{(1)})^{k-1}f$$

where $(\ )^{k-1}$ denotes the ordinary matrix power. One further example is for the tree

$t =$

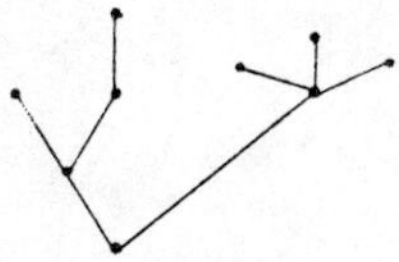

for which $F(t)(y_{n-1}) = f''(f''(f,f'(f)),f'''(f,f,f))$. For each tree we also define three numbers: $r(t)$, the number of nodes in t, $\sigma(t)$, the order of the symmetry group of t and $\gamma(t)$, the product of $r(u)$ over u where for each node of t, u is the subtree formed from that node and all nodes that can be reached from it by following upward growing branches. In the example given above we have $r(t) = 9$, $\sigma(t) = 6$ and $\gamma(t) = 288$. We also have $r(t_k)=k$, $\sigma(t_k) = (k-1)!$ and $\gamma(t_k) = k$; $r(t^k) = k$, $\sigma(t^k) = 1$ and $\gamma(t^k) = k!$, and $r(t_{k,l}) = k+l$, $\sigma(t_{k,l}) = (k-1)!\ (l-1)!$ (or if $l=1$, $\sigma(t_{k,l}) = k!$) and $\gamma(t_{k,l}) = l(k+l)$.

With this terminology, we can write down the formal Taylor series for $y(x_n)$ as follows

$$y(x_{n-1}+h) = y(x_{n-1}) + \sum_{t\in T} \frac{h^{r(t)}}{\sigma(t)\ \gamma(t)} F(t)(y(x_{n-1})). \qquad (2.5)$$

Thus, the first few terms of the expansion are, writing $y(x_{n-1})=y_{n-1}$,

$$y(x_{n-1}+h)=y_{n-1}+hf+\frac{1}{2}h^2f'(f)+h^3(\ \frac{1}{6}f''(f,f)+\ \frac{1}{6}f'(f'(f)))$$

$$+\ h^4(\ \frac{1}{24}f'''(f,f,f)+\frac{1}{8}f''(f,f'(f))+\ \frac{1}{24}f'(f''(f,f))+\frac{1}{24}f'(f'(f'(f))))+\ldots$$

For the computed solution, the particular coefficients a_{ij}, b_j clearly play a part. With each tree t we associate a polynomial $\Phi(t)$ in these quantities. To form $\Phi(t)$, attach labels i,j,k,... to each of the nodes of t, with i the label attached to the root, form the product of b_i and of a_{jk} for each upward growing branch from j to k, and then sum over each label from 1 to s. In writing formulae for $\Phi(t)$ we will always use the abreviation

$c_j = \sum_k a_{jk}$ already referred to. For the example of t given above we have

$$\Phi(t) = \sum_{i,j,k,l=1}^{s} b_i\, a_{ij}\, c_j\, a_{jk}\, c_k\, a_{il}\, c_l^3$$

while for the special types of trees t_k, t^k and $t_{k,l}$ we have

$$\Phi(t_k) = \sum_{i=1}^{s} b_i\, c_i^{k-1} \tag{2.6}$$

$$\Phi(t^k) = b^T A^{k-2} c \tag{2.7}$$

where A,b,c are the various arrays of coefficients and

$$\Phi(t_{k,l}) = \sum_{i,j=1}^{s} b_i c_i^{k-1} a_{ij} c_j^{l-1} \tag{2.8}$$

The formal Taylor series for the computed solution can now be written down

$$y_n = y_{n-1} + \sum_{t\in T} \frac{h^{r(t)}}{\sigma(t)} \Phi(t)F(t)\ (y_{n-1}) \tag{2.9}$$

and the first few terms are

$$y_n = y_{n-1} + h\sum b_i f + h^2 \sum b_i c_i f'(f) + h^3\left\{\frac{1}{2}\sum b_i c_i^2 f''(f,f) + \sum b_i a_{ij} f'(f'(f))\right\}$$

$$+ h^4\left\{\frac{1}{6}\sum b_i c_i^3 f'''(f,f,f) + \sum b_i c_i a_{ij} c_j f''(f,f'(f))\right.$$

$$\left. + \frac{1}{2}\sum b_i a_{ij} c_j^2\, f'(f''(f,f)) + \sum b_i a_{ij} a_{jk} c_k f'(f'(f'(f)))\right\}$$

$$+ \cdots$$

Comparing the expansions for y_n and $y(x_{n-1}+h)$ (equations (2.5) and (2.9)), term by term, we see that a method will be of order p if $\Phi(t) = 1/\gamma(t)$ for all t with no more than p nodes. Thus if p=4 we have the conditions,

$$\sum b_i = 1,\ \sum b_i c_i = \frac{1}{2},\ \sum b_i c_i^2 = \frac{1}{3},\ \sum b_i a_{ij} c_j = \frac{1}{6},\ \sum b_i c_i^3 = \frac{1}{4},$$

$$\sum b_i c_i a_{ij} c_j = \frac{1}{8},\ \sum b_i a_{ij} c_j^2 = \frac{1}{12},\ \sum b_i a_{ij} a_{jk} c_k = \frac{1}{24}.$$

In an explicit method, we wish to compute $Y_1, Y_2, \ldots, Y_s$ in turn, each depending only on the previously computed members of this sequence. Hence, for explicit methods, $a_{ij}=0$ unless $i>j$. It is easy to see that we cannot obtain order p with s stages unless $p \leqslant s$ because if t is chosen as the tree t^p, then the corresponding order condition is, from (2.7),

$$b^T A^{p-2} c = \frac{1}{p!} \tag{2.10}$$

If p>s, because of the strictly triangular structure of A and the fact that $c_1=0$, the left hand side of 2.10 is zero and thus 2.10 cannot be satisfied. It is also known that an order p cannot be obtained when p>4 unless s>p.

For an implicit s-stage Runge-Kutta method, the best possible order is 2s. The reason for this is considered in the next section.

The work in Section 2 has been reported in various places and in particular in Butcher (1963) and Butcher (1965a).

3. OPTIMAL ORDER IMPLICIT RUNGE-KUTTA METHODS

In considering s stage implicit Runge-Kutta methods, we make use of a number of propositions concerning the coefficients of the method. For any particular method, some of these propositions may be true and we shall derive some relationships between them. In each case, the proposition is given a name (A,B,...,G) for easy reference and these names together with the statements of the propositions are as follows:

A : a_{ij} satisfy $\sum_j a_{ij} c_j^{k-1} = \frac{1}{k} c_i^k$ for $i,k=1,2,\ldots,s$.

B : b_j satisfy $\sum_j b_j c_j^{k-1} = \frac{1}{k}$ for $k=1,2,\ldots,s$.

C : c_i are the zeros of the polynomial $P_s(2c-1)$, where P_s denotes the s degree Legendre polynomial.

D : down the jth column of the a_{ij} matrix, for each $j=1,2,\ldots,s$, it holds that

$$\sum_i b_i c_i^{k-1} a_{ij} = \frac{1}{k} b_j(1-c_j^k) \text{ for each } k=1,2,\ldots,s.$$

E : every pair of values $k,\ l=1,2,\ldots,s$ is such that

$$\sum_{i,j} b_i c_i^{k-1} a_{ij} c_j^{l-1} = \frac{1}{l(k+l)}.$$

F : for each $k=1,2,\ldots,2s$, it holds that $\sum_j b_j c_j^{k-1} = \frac{1}{k}$

G : good! For all t satisfying $r(t) \leq 2s$. The order condition $\Phi(t) = 1/\gamma(t)$ is satisfied.

The set of relationships that hold between A,B,...,G will be expressed in terms of a diagram in the statement of the theorem below. In interpreting this diagram, an arrow denotes implication while two or more arrows into a point and a single arrow out should be understood to represent an $\wedge$ (AND) gate. Thus, for example,

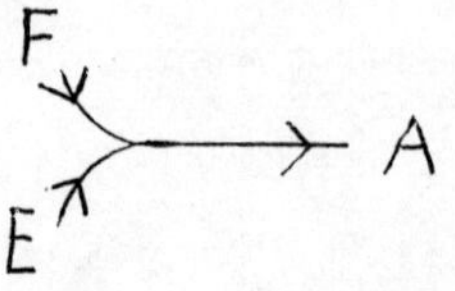

denotes the fact that if E and F are both true, then A is true.

THEOREM

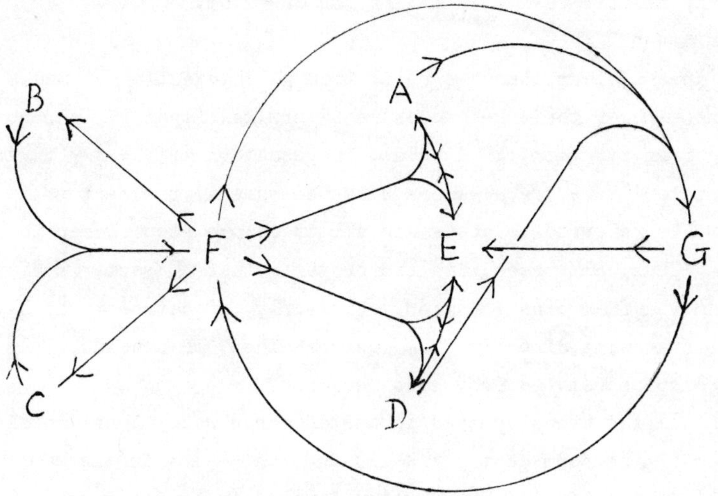

The truth of this theorem follows from a number of lemmas for which sketch proofs will be given

Lemma 1 $F \Rightarrow B$, $F \Rightarrow C$, $B \wedge C \Rightarrow F$.

This follows from the theory of Gauss-Legendre quadrature on the interval [0,1].

Lemma 2 If F then $A \Leftrightarrow E$.

For each $l=1,2,\ldots,s$, let $u_i = \sum_j a_{ij} c_j^{l-1} - \frac{1}{l} c_i^l$ for $i=1,2,\ldots,s$

and let $v_k = \sum_{i,j} b_i c_i^{k-1} a_{ij} c_j^{l-1} - \frac{1}{l(k+l)}$ for $k=1,2,\ldots,s$.

Given that F holds, it is easy to verify that $Mu=v$ where M is the matrix with k,i component equal to $b_i c_i^{k-1}$ $(i,k=1,2,\ldots,s)$, and because of the properties of Gauss-Legendre quadrature, M is non-singular. Therefore, if one of the vectors u, v is zero then so is the other.

<u>Lemma 3</u> If F then D <=> E.

This is proved in a similar way to Lemma 2.

<u>Lemma 4</u> G => F, G => E.

To prove F, substitute $t=t_k$ into $\Phi(t) = 1/\gamma(t)$ making use of (2.6).
To prove E, substitute $t=t_{k,1}$ making use of (2.8).

<u>Lemma 5</u> F $\wedge$ A $\wedge$ D => G

If t has a node, other than the root, from which exactly k-1 nodes branch and each of these k-1 nodes is a terminal (that is, has nothing branching from it) then, if A holds, the equation $\Phi(t) = 1/\gamma(t)$ is equivalent to $\Phi(u) = 1/\gamma(u)$ where u is the same as t except that the k-1 nodes referred to above are all moved one step closer to the root. Thus, all trees with the property that characterised t, can be removed from consideration. Similarly, if a tree t has k-1 terminals branching directly from its root then, if D holds, this tree can also be removed from consideration.

With all the trees removed in one of these ways, there remain only trees of the form of t_k ($k \leq 2s$) and, as we saw in the proof of Lemma 4, the equation $\Phi(t_k) = 1/\gamma(t_k)$ for $k=1,2,\ldots,2s$ is equivalent to F.

From lemmas 1 to 5, the theorem follows.

<u>COROLLARY</u> A $\wedge$ B $\wedge$ C <=> G

This follows from the diagram in the theorem by tracing through the various chains of implications. What this corollary means is that there is a unique method of order 2s for which the numbers $c_1, c_2, \ldots, c_s$ are determined by C, the numbers $b_1, b_2, \ldots, b_s$ are determined by B and, finally, $a_{11}, a_{12}, \ldots, a_{ss}$ are determined by A.

If s=1 we obtain what can be called the implicit mid-point rule

$$\begin{array}{c|c} \frac{1}{2} & \frac{1}{2} \\ \hline & 1 \end{array}$$

whereas if s=2 we obtain the method of Hammer and Hollingsworth (1955)

$$\begin{array}{c|cc}
\frac{1}{2} - \frac{\sqrt{3}}{6} & \frac{1}{4} & \frac{1}{4} - \frac{\sqrt{3}}{6} \\[2ex]
\frac{1}{2} + \frac{\sqrt{3}}{6} & \frac{1}{4} + \frac{\sqrt{3}}{6} & \frac{1}{4} \\[1ex]
\hline
 & \frac{1}{2} & \frac{1}{2}
\end{array}$$

In Butcher (1964), the general class of implicit Runge-Kutta methods was investigated and the methods of optimal order were given for s=3,4,5. The first of these is

$$\begin{array}{c|ccc}
\frac{1}{2} - \frac{\sqrt{15}}{10} & \frac{5}{36} & \frac{2}{9} - \frac{\sqrt{15}}{15} & \frac{5}{36} - \frac{\sqrt{15}}{30} \\[2ex]
\frac{1}{2} & \frac{5}{36} + \frac{\sqrt{15}}{24} & \frac{2}{9} & \frac{5}{36} - \frac{\sqrt{15}}{24} \\[2ex]
\frac{1}{2} + \frac{\sqrt{15}}{10} & \frac{5}{36} + \frac{\sqrt{15}}{30} & \frac{2}{9} + \frac{\sqrt{15}}{15} & \frac{5}{36} \\[1ex]
\hline
 & \frac{5}{18} & \frac{4}{9} & \frac{5}{18}
\end{array}$$

By modifying some of propositions A to F appropriately, it is an easy matter to derive other Runge-Kutta methods whose order is not much less than 2s but which have better stability properties than do the methods of optimal order. We examine the stability of Runge-Kutta methods in the next section.

4. STABILITY PROPERTIES

If we can consider the solution of the test equation $y'=qy$ and write $hq=z$, then (2.3), (2.4) can be written

$$(I - zA)Y = y_{n-1}\, e \qquad (4.1)$$

$$y_n = y_{n-1} + zb^T Y \qquad (4.2)$$

Where I is the s×s unit matrix, A and b are the matrix and vector of coefficients introduced in Section 2 and Y, e are the s dimensional vectors given by

$$Y = \begin{bmatrix} Y_1 \\ Y_2 \\ \cdot \\ \cdot \\ \cdot \\ Y_s \end{bmatrix}, \quad e = \begin{bmatrix} 1 \\ 1 \\ \cdot \\ \cdot \\ \cdot \\ 1 \end{bmatrix}$$

From 4.1 and 4.2 we find that $y_n = R(z)\, y_{n-1}$ where

$$R(z) = 1+zb^T(I-zA)^{-1}e \qquad (4.3)$$

A complex number z will be in the stability region for this Runge-Kutta method if $|R(z)| \leq 1$ (or $|R(z)| < 1$ in some definitions of the stability region; the two definitions are related in that the first type of stability region is the closure of the second and the second is the interior of the first).

Since (I-zA) is an s×s matrix, its determinant is a polynomial in z of degree (not exceeding) s and since its cofactors are (s-1)×(s-1) matrices, these in turn are polynomials of degree (not exceeding) s-1. Thus R(z) is a rational function of degrees (not exceeding) s for both numerator and denominator.

Expanding the factor $(I-zA)^{-1}$ as a series $I+zA+z^2A^2+\ldots$ and noting that $b^TA^{k-1}e$ is $\Phi(t^k) = 1/k!$ as long as $k \leq p$, the order of the method, we see that

$$R(z) = e^z+O(z^{p+1})$$

as $z \to 0$. If p=2s, we see that for the optimal order methods considered in Section 3, R(z) is the (s,s) Padé approximation to e^z. As well as methods with this property, we will also study the stability region for methods in which R(z) is one of the first two subdiagonals of the Padé table.

Thus, we suppose $R(z) = P(z)/Q(z)$ where Q is a polynomial of degree s, P is a polynomial of degree s-d (d=0, 1 or 2) and $R(z) = e^z+O(z^{2s-d+1})$. As a step in determining the stability region for these methods, we prove

<u>Theorem</u> For R(z) a diagonal, or one of the first two subdiagonals, of the Pade table for e^z then for all z on the imaginary axis, $|R(z)| \leq 1$. Since $P(z) = Q(z)e^z+O(z^{2s-d+1})$ as $|z| \to 0$, we have for real y,

$$|Q(iy)|^2 - |P(iy)|^2 = Q(iy)Q(-iy)-P(iy)P(-iy)$$
$$= O(y^{2s-d+1})$$

But $Q(iy)Q(-iy)-P(iy)P(-iy)$ is an even polynomial in y of degree no more than 2s and is, therefore, equal to zero (in which case $|R(iy)|=1$) when d=0 or of the form cy^{2s} for some real c when d=1 or 2. In the last case, the fact that P has degree less than s guarantees that $c\geq 0$ and we find $|R(iy)| \leq 1$ (in fact $c>0$ and hence, for $y>0$, $|R(iy)| < 1$).

An investigation of the zeros of Q(z) for these methods establishes that they are all in the right half plane. Combining this fact with the behaviour we have established on the imaginary axis and the fact that $|R(z)| \to 1$ (for the diagonal case) or that $|R(z)| \to 0$ (for the subdiagonal cases) as $|z| \to \infty$ leads to the following result which is stated without further proof.

<u>Corollary</u> Methods which lead to a member of the diagonal or one of the two main diagonals of the Padé table, are A-stable.

The fact that $|R(z)| \to 0$ for the subdiagonal methods makes them preferable for stiff problems. Methods of this type have been investigated in detail by

Ehle (1969b), Chipman (1971b) and Axelsson (1969).

5. IMPLEMENTATIONS OF IMPLICIT RUNGE-KUTTA METHODS

The obvious iterative scheme for evaluating $Y_1, Y_2, \ldots, Y_s$ is not appropriate for use with stiff problems and it becomes necessary to use the Newton-Raphson method or some modification of it. For the full Newton-Raphson scheme, if $Y_1+\delta_1$, $Y_2+\delta_2, \ldots, Y_s+\delta_s$ are the values of the modified approximations after a single iteration, then $\delta_1, \delta_2, \ldots, \delta_s$ are given by the linear system

$$\begin{bmatrix} I - ha_{11}J_1 & -ha_{12}J_2 & \cdots & -ha_{1s}J_s \\ -ha_{21}J_1 & I - ha_{22}J_2 & \cdots & -ha_{2s}J_s \\ \cdot & \cdot & \cdots & \cdot \\ \cdot & \cdot & \cdots & \cdot \\ \cdot & \cdot & \cdots & \cdot \\ -ha_{s1}J_1 & -ha_{s2}J_2 & \cdots & I - ha_{ss}J_s \end{bmatrix} \begin{bmatrix} \delta_1 \\ \delta_2 \\ \cdot \\ \cdot \\ \cdot \\ \delta_s \end{bmatrix} + \begin{bmatrix} Y_1 - y_{n-1} - h\sum a_{1j}F_j \\ Y_2 - y_{n-1} - h\sum a_{2j}F_j \\ \cdot \\ \cdot \\ \cdot \\ Y_s - y_{n-1} - h\sum a_{sj}F_j \end{bmatrix} = 0 \qquad (5.1)$$

where $F_1=f(Y_1), F_2=f(Y_2), \ldots, F_s=f(Y_s)$ and $J_1=f'(Y_1)$, $J_2=f'(Y_2), \ldots$, $J_s=f'(Y_s)$.

In typical programs, $J_1, J_2, \ldots, J_s$ are not altered after each iteration so that the inverse (or, more efficiently, the L U factorisation) of the matrix in (5.1) need be computed less frequently.

An enormous gain is obtained in computational efficiency in these iterations if $a_{ij}=0$ for $i<j$ and $a_{11}, a_{22}, \ldots, a_{ss}$ are all equal. If $J_1, J_2, \ldots, J_s$ are approximated by f' evaluated as some fixed point (say $J=J_1=J_2=\ldots=J_s$) then only the single block inversion (or L U factorization) for the matrix $I-ha_{11}J$ is necessary. We will deal briefly in Section 6 with these "semi-explicit" methods, together with the class of method proposed by Rosenbrock (1963) which are essentially equivalent to just a single iteration of an implicit Runge-Kutta method.

6. SEMI-EXPLICIT AND ROSENBROCK METHODS

For a semi-explicit method, $a_{ij}=0$ unless $i \geq j$. Since these methods are implicit it is possible that amongst them can be found methods with good stability properties. As mentioned above they can be implemented more efficiently than general implicit methods.

Although these methods were proposed as long ago as in Butcher (1964), apart from isolated examples for example in the work of Ehle (1969b) and Chipman (1971b), no systematic study of them seems to have been carried out until the work of Nørsett (1974). In this study the special case where a_{ij} is the same for all i was considered. For this special case, it has been proved that an order s+2 cannot be achieved for any s stage methods.

We now consider the case s=2 and we attempt to find a 3rd order method with $a_{11}=a_{22}$. Thus we must satisfy the equations

$$b_1 + b_2 = 1 \tag{6.1}$$

$$b_1c_1+b_2c_2= \frac{1}{2} \tag{6.2}$$

$$b_1c_1^2+b_2c_2^2= \frac{1}{3} \tag{6.3}$$

$$b_1a_{11}c_1+b_2a_{21}c_1+b_2a_{22}c_2= \frac{1}{6} \tag{6.4}$$

Eliminating b_1 from (6.1) we find

$$b_2(c_2-c_1) = \frac{1}{2} - c_1 \tag{6.5}$$

Multiplying (6.1) by c_1c_2, (6.2) by $-(c_1+c_2)$ and adding the sum of these products to (6.3) gives

$$c_1c_2 - \frac{1}{2}(c_1+c_2) + \frac{1}{3} = 0 \tag{6.6}$$

Using (6.2), the facts that $a_{22}=a_{11}=c_1$ and $a_{21}=c_2-c_1$ enables (6.4) to be rewritten as

$$b_2(c_2-c_1)c_1 = \frac{1}{6} - \frac{c_1}{2} \tag{6.7}$$

We now substitute from (6.5) into (6.7) to find that c_1 must satisfy

$$6c_1^2 - 6c_1 + 1 = 0 \tag{6.8}$$

Using (6.8), (6.6), (6.7) and (6.1) to find in turn the values of c_1, c_2, b_2 and b_1 and inserting, finally, the values of a_{11}, a_{12}, a_{22} leads to the following method or a similar method with $\sqrt{3}$ replaced by $-\sqrt{3}$

$$\begin{array}{c|cc} \frac{3+\sqrt{3}}{6} & \frac{3+\sqrt{3}}{6} & 0 \\ \frac{3-\sqrt{3}}{6} & -\frac{\sqrt{3}}{3} & \frac{3+\sqrt{3}}{6} \\ \hline & \frac{1}{2} & \frac{1}{2} \end{array}$$

The function $R(z)$ associated with the stability analysis of this method is given by

$$R(z) = \frac{1 - \frac{\sqrt{3}}{3} z - \frac{1+\sqrt{3}}{6} z^2}{1 - \frac{3+\sqrt{3}}{3} z + \frac{2+\sqrt{3}}{6} z^2}$$

so that $|R(z)| \to \frac{1+\sqrt{3}}{2+\sqrt{3}} < 1$ as $|z| \to \infty$ and an argument similar to that in the theorem of Section 4 establishes the fact that this method is A-stable. Note that in the method where the sign of $\sqrt{3}$ is changed, the stability region is bounded.

A number of special methods are given in Nørsett's paper and these are, in general, supplied with local error estimates for step size control.

The methods of Rosenbrock (1962), which were also studied, for example, by Allen (1969) and Haines (1969), take a form essentially the same as a single iteration of a semi explicit method. Thus, if coefficients a_{ij}, d_{ij} $(1 \leqslant j < i \leqslant s)$ and θ_i, b_i $(1 \leqslant i \leqslant s)$ are given, then from a given solution value y_{n-1} at the beginning of step number n we compute the vectors $Y_1, Y_2, \ldots, Y_s$, $Z_1, Z_2, \ldots, Z_s$, $F_1, F_2, \ldots, F_s$, the matrices $J_1, J_2, \ldots, J_s$ and the results after the step, y_n, by the equations

$$Y_i = y_{n-1} + h \sum_j a_{ij} F_j \qquad (i=1,2,\ldots,s)$$

$$Z_i = y_{n-1} + h \sum_j d_{ij} F_j \qquad (i=1,2,\dots,s)$$

$$F_i = (I - h\theta_i J_i)^{-1} f(Y_i) \qquad (i=1,2,\dots,s)$$

$$J_i = f'(Z_i) \qquad (i=1,2,\dots,s)$$

$$y_n = y_{n-1} + h \sum_j b_j F_j$$

With such methods, it is possible to obtain A-stability without implicitness. A particular method from Calahan (1967), has the property that $d_{ij}=0$ for all i, j and θ_i is constant for all i. Thus only one Jacobian evaluation is required per step. This two stage method has $\theta_1=\theta_2 = \frac{3+\sqrt{3}}{6}$, $a_{21} = -2\sqrt{3}/3$, $b_1 = 3/4$, $b_2 = 1/4$ and is A-stable with order 3.

11

MULTISTEP METHODS FOR STIFF PROBLEMS

In this chapter we shall first discuss briefly some of the implicit linear multistep methods suitable for solving stiff initial-value problems and shall describe in some detail the variable-step, variable-order implementations of the stiffly stable backward differentiation methods made by Gear (1971a) and by Brayton, Gustavson and Hachtel (1972). A number of stable modified multistep methods are then shown to belong to the class of generalized multistep methods with variable coefficients studied by Lambert and Sigurdsson (1972). We conclude with a description of the class of implicit k-step second-derivative methods proposed by Enright (1974).

1. IMPLICIT LINEAR MULTISTEP METHODS

In discussing the stability requirements of methods for stiff problems in Chapter 9, we indicated that for a linear multistep method

$$\sum_{j=0}^{k} \alpha_j \, y_{n+j} - h \sum_{j=0}^{k} \beta_j \, f(x_{n+j}, y_{n+j}) = 0, \quad \alpha_k = 1 , \tag{1.1}$$

to possess any $A(\alpha)$-stability or stiff stability property, it must necessarily be "implicit", i.e. $\beta_k \neq 0$. This follows directly from the root condition for absolute stability on the associated stability polynomial

$$\rho(\xi) - h\lambda . \sigma(\xi) = 0 \tag{1.2}$$

since for $\beta_k = 0$, one root is unbounded as $\mathrm{Re}(h\lambda) \to -\infty$.

Dahlquist (1963) has proved that the maximum order of A-stable linear multistep methods is two, and that of such methods the formula with the smallest error constant $C = \frac{1}{12}$ is the one-step trapezium rule

$$y_{n+1} - y_n - \tfrac{1}{2}h\left\{f(x_n,y_n) + f(x_{n+1},y_{n+1})\right\} = 0 . \tag{1.3}$$

The use of the trapezium rule for solving stiff problems will be discussed in the following chapter.

The families of A-stable linear k-step methods of order p = k are given by

$$y_{n+1} - y_n - h\left\{(1-\theta)\, f_{n+1} + \theta f_n\right\} = 0, \quad \theta<\tfrac{1}{2}\ , \tag{1.4}$$

and

$$y_{n+2} - (1+a)\, y_{n+1} + a\, y_n$$
$$- h\left\{[\tfrac{1}{2}(1+a) + \theta]\, f_{n+2} + [\tfrac{1}{2}(1-3a) - 2\theta]\, f_{n+1} + \theta f_n\right\} = 0, \tag{1.5}$$

with $-1 < a < 1$ and $a + 2\theta > 0$ (Liniger, 1968; Brunner, 1972). Methods based on formula (1.4) have been implemented by Liniger and Willoughby (1970), using exponential fitting to determine θ, and by Prothero and Robinson (1974) using fixed $\theta = 0.45$. A method is said to be exponentially fitted at a value $\bar{\lambda}$ if, when applied to the equation $y' - \lambda y = 0$ with exact initial conditions, it gives an exact solution for $\lambda = \bar{\lambda}$. Thus, in applying formula (1.4), θ is chosen so that

$$y_{n+1} = \left\{\frac{1+\theta h\bar{\lambda}}{1-(1-\theta)h\bar{\lambda}}\right\} y_n = e^{h\bar{\lambda}}\, y_n\ .$$

Exponential fitting using formula (1.5) has also been studied (Liniger, 1969).

Turning now to consider methods that are $A(\alpha)$-stable or stiffly stable, we note that stiff stability implies $A(\alpha)$-stability for $0 < \alpha < \mathrm{Tan}^{-1}(\theta/D)$. Widlund (1967) has shown that $A(\alpha)$-stable linear multistep methods exist for any $\alpha \in [0, \frac{\pi}{2})$ with $k = p = 3$ and $k = p = 4$. Stiffly stable methods up to order eleven have been found (Gear, 1969; Dill and Gear, 1971; Jain and Srivastava, 1970), while Cryer (1973) has proved the existence of A_0- stable (absolutely stable for all real

$h\lambda < 0$ - not equivalent to A(0)-stability) linear multistep methods of arbitrarily high order; further results on the A_0-stability of k-step methods of order p have been obtained by Liniger (1975).

2. BACKWARD DIFFERENTIATION FORMULAE

Easily the most widely used class of linear multistep methods for stiff problems are the implicit "backward differentiation formulae". As the name suggests, the derivative at the current point, f_{n+k}, is approximated using the (k+1) values y_{n+j} $(0 \leq j \leq k)$ giving linear multistep methods of the form

$$\sum_{j=0}^{k} \alpha_j \, y_{n+j} - h\beta_k f_{n+k} = 0, \; \alpha_k = 1 \; . \tag{2.1}$$

For k = 1, we have the Euler backward difference method. These formulae, with k = 1 and 2, were first proposed for solving stiff problems by Curtiss and Hirschfelder (1952). Gear (1969) showed that the methods (2.1) for k = 1,2,...,6 satisfied his definition of stiff stability, and he outlined a sophisticated implementation of the methods as an algorithm with automatic step-size and order changing. This implementation was subsequently published (coupled with the Adams-Bashforth-Moulton methods for non-stiff systems) as a FORTRAN subroutine DIFSUB (Gear, 1971a).

The regions of absolute stability for the methods (2.1) are shown in Figure 1. For k = 1,2 the methods are A-stable, while for k = 3,...,6, the methods are stiffly stable and hence A(α)-stable for some $\alpha \in (0, \frac{\pi}{2})$. The maximum values α_{max} for which the methods are A(α)-stable have been calculated using a criterion for A(α)-stability derived by Norsett (1969a).

In implementing backward differentiation formulae, Gear (1971a) has used the Nordsieck (1962) representation (described in Chapter 6), in which the (k+1) quantities occurring in the k-step method are stored at each step as the elements of the Nordsieck vector

$$(y_n, h_n y_n', \tfrac{1}{2}h_n^2 y_n'', \ldots, \frac{1}{k!} h_n^k y_n^{(k)}) \; .$$

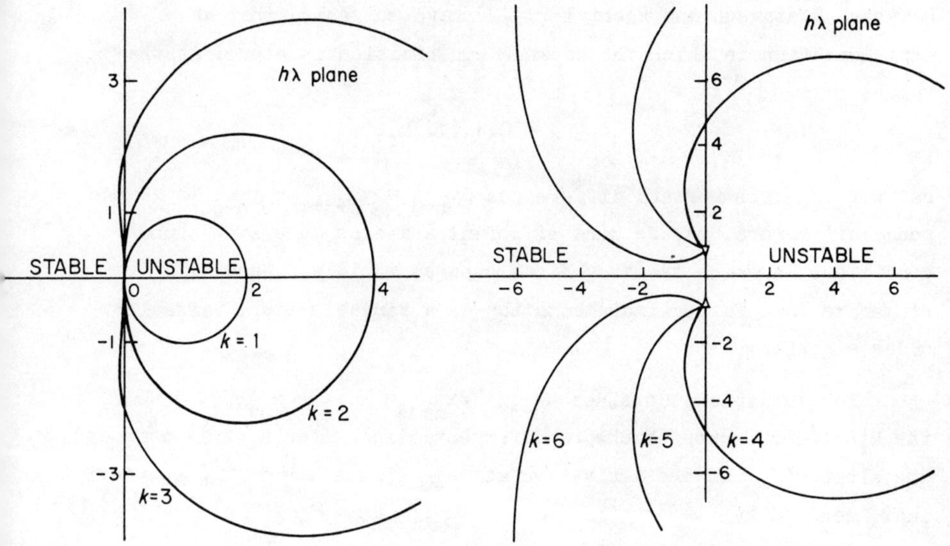

FIG. 1 — Regions of absolute stability for the k-step backward differentiation methods, k = 1, 2...... 6 (Gear, 1969)

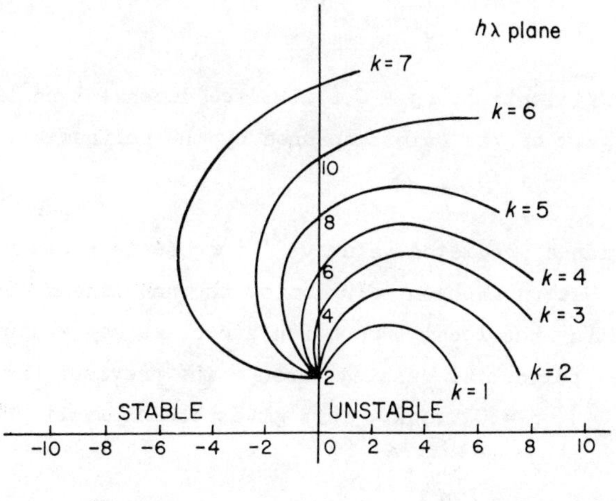

FIG. 2 — Regions of absolute stability for the k-step second-derivative methods of Enright (1974), k = 1, 2..... 7

Brayton, Gustavson and Hachtel (1972) have suggested that an implementation in which the backward information is stored as the values

$$y_{n+j} \quad (j = 0,1,\ldots,k),$$

but working with forward differences $\Delta y_{n+j} = y_{n+j+1} - y_{n+j}$ to reduce round-off errors, may be more efficient and also more stable under conditions in which the step sizes changes rapidly. We will summarize this latter implementation as a variable-step, variable-order algorithm.

For variable step sizes $h_{n+j} = \nabla x_{n+j} = x_{n+j} - x_{n+j-1}$, $j = 0,\ldots,k$ the kth-order, k-step backward differentiation formula (2.1) defines the slope y'_{n+k} as the derivative at x_{n+k} of the kth-degree polynomial in x, denoted by

$$P\ (k,\ n+k,\ x) \quad ,$$

which passes through the k+1 points (x_{n+j}, y_{n+j}), $j = 0,1,\ldots,k$. Thus

$$\begin{aligned} y'_{n+k} &= P'(k,\ n+k,\ x_{n+k}) \\ &= -\frac{1}{h_{n+k}} \sum_{j=0}^{k} \delta_j\, y_{n+j} = f(x_{n+k}, y_{n+k}) \ , \end{aligned} \tag{2.2}$$

where the coefficients δ_j $(j = 0,\ldots,k)$ are dependent on the spacing h_{n+j}, $j = 1,\ldots,k$ of the points spanned by the polynomial. (For constant h, $\delta_j \equiv -\alpha_j/\beta_k$).

To provide a predicted value $y^{(0)}_{n+k}$ for use as a starting value in the iterative Newton-Raphson solution of the non-linear equations (2.2), and in estimating the local truncation error, we may evaluate at x_{n+k} the kth-degree polynomial passing through the previous (k+1) points (x_{n+j-1}, y_{n+j-1}), $j = 0,\ldots,k$. This gives the explicit predictor

$$\begin{aligned} y^{(0)}_{n+k} &= P(k,\ n+k-1,\ x_{n+k}) \\ &= \sum_{j=0}^{k} \gamma_j\, y_{n+j-1} \ , \end{aligned} \tag{2.3}$$

where the coefficients γ_j, $j = 0,1,\ldots,k$ are dependent on h_{n+j}, $j = 0, \ldots, k$.

The coefficients δ_j, γ_j, $j = 0, \ldots, k$ may be determined at each step n+k by requiring that equations (2.2) and (2.3) be accurate for all polynomials $y = p(x)$ of degree $\leq k$. This condition gives the sets of linear equations

$$H_k(n, x_{n+k})\,\delta = e_2 \quad , \tag{2.4}$$

and

$$H_k(n-1, x_{n+k})\,\gamma = e_1 \, , \tag{2.5}$$

where $\delta^T = (\delta_0, \delta_1, \ldots, \delta_k)$, $\gamma^T = (\gamma_0, \gamma_1, \ldots, \gamma_k)$, $e_1^T = (1, 0, \ldots, 0)$, $e_2^T = (0, 1, \ldots, 0)$, and H_k is a Vandermonde matrix with

$$[H_k(m, x_{n+k})]_{ij} = \left(\frac{x_{n+k} - x_{m+j}}{h_{n+k}}\right)^i , \ (i, j = 0, 1, \ldots, k). \tag{2.6}$$

The coefficients δ_j, γ_j may be evaluated at each step in roughly 10(k+1) operations.

Brayton et al (1972) show that the principal local discretization error, of the form

$$C^*_{k+1}\, h^k_{n+k}\, y^{(k+1)}\,(x_{n+k}) \; ,$$

may be expressed as the difference in the predicted and corrected values

$$d_{n+k} = \frac{1}{x_{n+k} - x_{n-1}}\,(y_{n+k} - y^{(0)}_{n+k}) + O\,(h^{k+1}_{n+k}) \, .$$

While Gear (1971a) controls the step size by restricting the local truncation error (defined as $h_{n+k}\, d_{n+k}$) to a constant (relative) error limit per integration step, Brayton et al propose an error limit per unit length of integration range.

The technique used for changing step size in Gear's implementation is equivalent to interpolation through saved values with spacing h_{old} to obtain values with spacing h_{new} for substitution in the constant-step formula. This approach has been shown (Brayton et al, 1972; Gear and Tu, 1974), to be less stable than the direct use of multistep formulae based on unequal intervals, and in Gear's algorithm the step is held constant for k+1 steps after a step or order change. The variable-step approach of Brayton et al with the size adjusted at every step, should reduce the number of integration steps required and provide a significant saving over Gear's implementation for large systems of equations.

The test for changing order in both the Gear and Brayton implementations is only made after k+1 steps at given order k, and requires estimates of the local truncation errors for the backward differentiation formulae of order k-1, k and k+1. These estimates require the additional calculation, in the variable-step approach, of $x_{n+k}^{(0)}$ predicted by the k-1 and k+1-step predictors. Gear and Watanabe (1974) have shown that the variable-step approach appears to be more stable to changes in order than the interpolation technique.

An implementation of the variable-step backward differentiation formulae (2.2), based on the Nordsieck vector of calculated values, has recently been published by Byrne and Hindmarsh (1975).

3. GENERALIZED MULTISTEP METHODS

The difficulties that are caused by the necessarily implicit nature of the stiffly stable linear multistep methods have prompted several investigations into possible ways of stabilizing explicit multistep methods by using the Jacobian matrix. These various approaches are best described and compared by considering the following class of generalized multistep methods (Lambert and Sigurdsson, 1972)

$$\sum_{j=0}^{k}\left\{\alpha_j^{(0)} I + \sum_{m=1}^{M} (-1)^m \alpha_j^{(m)} h^m J_n^m\right\} y_{n+j}$$

$$- h\sum_{j=0}^{k}\left\{\beta_j^{(0)} I + \sum_{m=1}^{M-1} (-1)^m \beta_j^{(m)} h^m J_n^m\right\} f_{n+j} = 0,\ \alpha_k^{(0)} = 1. \quad (3.1)$$

The $\alpha_j^{(m)}$, $\beta_j^{(m)}$ are scalar coefficients and J_n is an approximation to the Jacobian matrix $\partial f/\partial y$ evaluated at or near x_{n+k}. If the coefficients $\beta_k^{(m)}$, m = 0,1,...,M - 1, are zero, equation (3.1) gives y_{n+k} as the solution to a set of linear equations of the form

$$\left\{\alpha_k^{(0)} I + \sum_{m=1}^{M} \alpha_k^{(m)} h^m J_n^m\right\} y_{n+k} - g = 0 , \tag{3.2}$$

where g is known, and the method is said to be "linearly implicit". Such linearly implicit methods would appear to offer a practical alternative to fully-implicit methods.

The order p of a method (3.1) is independent of the accuracy of J_n, although the truncation error has the more complex form

$$t_{n+k} = h^{p+1} \sum_{j=0}^{\text{Min}(p+1,M)} C_{p+1-j}^{(j)} J_n^j y^{(p+1-j)}(x_{n+k}) + O(h^{p+2}) . \tag{3.3}$$

Lambert and Sigurdsson (1972) show that the conditions for order p ⩾ M are

$$\sum_{j=0}^{k} \alpha_j^{(m)} = 0 , \qquad m = 0,1,\ldots,M;$$

and

$$\sum_{j=0}^{k} j^{t-1}\left\{j\alpha_j^{(m)} - t\beta_j^{(m)}\right\} = 0 , \qquad t = 1, 2,\ldots,p-m , \quad m = 0, 1,\ldots,M .$$

If the roots of the polynomial

$$\rho(\xi) = \sum_{j=0}^{k} \alpha_j^{(0)} \xi^j , \tag{3.4}$$

are $\xi_1, \xi_2, \ldots, \xi_k$, with principal root $\xi_1 = 1$, the usual root condition for zero stability requires that $|\xi_r| \leq 1$, r = 1,2,...,k and that roots of unit modulus are simple. The region of absolute stability of a method (3.1) is that region of the complex plane $\bar{h} = h\lambda$ for which the roots ξ of the characteristic polynomial

$$\sum_{j=0}^{k}\left\{\alpha_j^{(0)} + \sum_{m=1}^{M}(\alpha_j^{(m)} - \beta_j^{(m-1)})\,\bar{h}^{m}\right\}\xi^j = 0 \qquad (3.5)$$

lie in the open unit circle.

Lambert and Sigurdsson (1972) restrict their study of methods (3.1) to "stabilized" methods, for which the parasitic roots $\xi_2,\ldots,\xi_k$ of equation (3.4) lie inside the unit circle, and are also roots of equation (3.5) for all $\bar{h}$. This is achieved by the stabilizing conditions

$$\sum_{j=0}^{k}\gamma_j^{(m)}\,\xi_r^j = 0\ , \qquad m = 1, 2,\ldots,M:\ r = 2, 3,\ldots,k\ ,$$

for distinct roots ξ_r of equation (3.4), where

$$\gamma_j^{(m)} = \alpha_j^{(m)} - \beta_j^{(m-1)} .$$

Corresponding conditions are obtained for multiple roots.
Thus the absolute stability region of "stabilized" methods (3.1) is governed only by the principal root $\xi_1(\bar{h})$ of equation (3.5), which for methods of order $p \geqslant M$ is shown to be given by

$$\xi_1(\bar{h}) = \frac{\displaystyle\sum_{m=0}^{M}\left\{\sum_{l=0}^{m}\frac{(-1)^l}{l!}\,\gamma_k^{(m-l)}\right\}\bar{h}^{m}}{\displaystyle\sum_{m=0}^{M}\gamma_k^{(m)}\,\bar{h}^{m}}$$

where $\gamma_k^{(0)} = \alpha_k^{(0)} = 1$. If $\xi_1(\bar{h})$ is an A-acceptable rational approximation to the exponential $e^{\bar{h}}$, the stabilized method (3.1) is A-stable. The maximum order of such methods is shown to be 2M. Some examples of linearly implicit stabilized methods (3.1) of orders 2, 3 and 4 have been derived by Lambert and Sigurdsson (1972). Van der Houwen and Verwer (1974) have studied the subset of stabilized methods having zero parasitic roots $\xi_2,\ldots,\xi_k$. The A-stable modification of the

Adams-Bashforth methods developed by Norsett (1969b), which gives methods of the form

$$y_{n+k} - e^{hJ} . y_{n+k-1} - h \sum_{j=0}^{k-1} c_j \left\{ f_{n+j} - J.y_{n+j} \right\} = 0 ,$$

form a class of stabilized methods with zero parasitic roots when a rational approximation is substituted for the exponential term. Similar methods of higher-order accuracy have also been proposed by Jain (1972).

4. SECOND DERIVATIVE METHODS

One means of making use of the Jacobian matrix in generalizing linear multistep methods is in the calculation of second derivatives y''. For an autonomous system of equations

$$y' - f(y) = 0 ,$$

we have

$$y'' - (\partial f/\partial y). \ y'=0 ,$$

so that we might consider using second-derivative multistep formulae of the form

$$\sum_{j=0}^{k} \left\{ \alpha_j \ y_{n+j} - h\beta_j \ y'_{n+j} - h^2 \gamma_j \ y''_{n+j} \right\} = 0 \ ; \ \alpha_k = 1 . \qquad (4.1)$$

Families of one- and two-step A-stable methods of this form have been proposed by Liniger and Willoughby (1970) and Jackson and Kenue (1974) respectively, with a free parameter to allow "exponential fitting". The methods obtained from these families by "exponential fitting at infinity" are the first two members of the class of stiffly stable k-step second-derivative methods, of order k+2, that have been developed by Enright (1974) and tested with promising results.

The region of absolute stability in the $\bar{h} = h\lambda$ plane of the formula (4.1) is that region for which the roots of the characteristic polynomial

$$\sum_{j=0}^{k} (\alpha_j - \beta_j \ \bar{h} - \gamma_j \ \bar{h}^2) \ \xi^j = 0 , \qquad (4.2)$$

lie in the open unit circle. The parasitic roots at the origin are set to zero in Enright's methods by choosing, as in the Adams methods,

$$\alpha_k = 1; \; \alpha_{k-1} = -1; \; \alpha_{k-2}, \ldots, \alpha_0 = 0 \; .$$

Similarly, stability at infinity is ensured by taking

$$\gamma_{k-1} = \gamma_{k-2}, \ldots, = \gamma_0 = 0; \; \gamma_k \neq 0 \, ,$$

so that all roots tend to zero as $|h\lambda| \to \infty$. The remaining (k+2) parameters γ_k and $\beta_0 \ldots \beta_k$ are chosen to maximize the order, so that we have the implicit formulae

$$y_{n+k} - y_{n+k-1} - h \sum_{j=0}^{k} \beta_j \, y'_{n+j} - h^2 \gamma_k \, y''_{n+k} = 0 \, . \qquad (4.3)$$

Enright has shown that these methods are stiffly stable for k = 1,2,..,7, and that k = 8 is not stiffly stable. The stability regions are shown in Figure 2.

A modified Newton-Raphson iteration is needed to solve the implicit sets of (non-linear) equations at each step. Enright uses the following iteration :

$$W_{n+k} \, (y_{n+k}^{(m+1)} - y_{n+k}^{(m)}) = -y_{n+k}^{(m)} + h\beta_k \, y'^{(m)}_{n+k} + h^2 \gamma_k \, y''^{(m)}_{n+k} + y_{n+k-1} + \sum_{j=0}^{k-1} \beta_j \, y'_{n+j},$$

$$m = 0, 1, \ldots , \qquad (4.4)$$

where

$$W_{n+k} \approx [I - h\beta_k \, \partial f/\partial y - h^2 \gamma_k \, (\partial f/\partial y)^2] \, .$$

The matrix W_{n+k} is only recalculated when necessary for convergence, or when the step size or order is changed. Note that for non-linear problems, calculation of the second derivative ,

$$y''^{(m)}_{n+k} = \frac{\partial f}{\partial y} \, (y_{n+k}^{(m)}) \, y'^{(m)}_{n+k} \; , \qquad (4.5)$$

requires the re-evaluation of the Jacobian matrix at each and every iteration. Also in the definition of the iteration matrix W_{n+k}, the variation $\frac{\partial}{\partial y}$ $(\partial f/\partial y).y'$ is neglected.

While Enright's methods show improvements in accuracy and in stability properties compared with the backward differentiation methods, the need to re-evaluate the Jacobian at each iteration of every step must make the methods expensive for solving large, non-linear systems of differential equations. If the matrix is not re-evaluated, however, so that, for example,

$$y''^{(m)}_{n+k} \approx \frac{\partial f}{\partial y}(y^{(0)}_{n+k})\, y'^{(m)}_{n+k} ,$$

the methods cease to be true second-derivative methods and become generalized methods of the type considered in the previous section.

12

EXTRAPOLATION METHODS FOR STIFF SYSTEMS AND A COMPARISON OF METHODS FOR STIFF PROBLEMS

1. INTRODUCTION

This chapter has two distinct parts. First, we describe an extrapolation method for stiff systems of ordinary differential equations. This method is not closely related to the Gragg-Bulirsch-Stoer method for non-stiff problems and requires some entirely new ideas.

In the second part of the chapter we describe the structure of some of the different types of stiff problems which occur in practice. Finally, we discuss some recent quantitative comparisons of current algorithms.

2. EXTRAPOLATION METHODS FOR STIFF ORDINARY DIFFERENTIAL EQUATIONS

Consider the trapezium rule

$$\underline{y}_{n+1} = \underline{y}_n + \frac{h}{2}(\underline{f}(x_n, \underline{y}_n) + \underline{f}(x_{n+1}, \underline{y}_{n+1})) \tag{2.1}$$

for finding a numerical solution to the problem

$$\underline{y}' = \underline{f}(x,\underline{y}), \quad \underline{y}(x_0) = \underline{\eta} \tag{2.2}$$

at the points $x_0, x_1, \ldots, x_N$ where $x_k = x_0 + kh$. It has been shown (Lindberg 1971a) that if $y \in C^{2M+1}[x_0, x_N]$ and

$$\underline{y}_0 = \underline{y}(x_0) + \sum_{i=1}^{M} \underline{c}_i h^{2i} + O(h^{2M+1}) \tag{2.3}$$

then

$$\underline{y}_k = \underline{y}(x_k) + \sum_{i=1}^{M} \underline{c}_i(x_k) h^{2i} + O(h^{2M+1}), \tag{2.4}$$

where $\underline{c}_i(x) \in C^{2M+1-2i}[x_0, x_N]$. In fact, it is possible to derive both differential and difference equations (Dahlquist and Lindberg 1973) for the functions $\underline{c}_i(x)$. For example, $\underline{c}_1(x)$ satisfies

$$\underline{c}_1' - \frac{\partial f}{\partial \underline{y}} \underline{c}_1 = \underline{y}'''/12, \quad \underline{c}_1(x_0) = \underline{c}_1. \tag{2.5}$$

The power series expansion (2.4) forms the basis for the methods of h^2-extrapolation and deferred correction. In the deferred correction method an approximation is calculated for the vector functions $\underline{c}_1(x)$, $\underline{c}_2(x)$,... in turn, and then the series (2.4) is used to improve the approximation $\underline{y}_k$ to $\underline{y}(x_k)$ for each value k. The method has received little attention in the current context. In the h^2-extrapolation method, for any point $t \in \{x_i\}$, we calculate $\underline{y}(t, h)$ and $\underline{y}(t, h/2)$ from (2.1) with step h and h/2 respectively, then from the series (2.4) we see that if

$$\overline{\underline{y}}(t, h) = \underline{y}(t, h/2) + (\underline{y}(t, h/2) - \underline{y}(t, h))/3, \tag{2.6}$$

then

$$\overline{\underline{y}}(t, h) = \underline{y}(t) + \sum_{i=2}^{M} \underline{d}_i(t)h^{2i} + O(h^{2M+1}) \tag{2.7}$$

for some vector functions $\underline{d}_i(x) \in C^{2M+1-2i}[x_0, x_N]$. A similar process can be devised to eliminate successive powers of h^2 on the right hand side of equation (2.7) hence giving an approximate solution of any required order of accuracy if the solution $\underline{y}$ is sufficiently smooth. The relations (2.6) and (2.7) are the foundation for extrapolation methods for stiff systems of equations.

In the program IMPEX described in Lindberg (1972a) and updated in Lindberg (1973), the trapezium rule is not used; but instead a similar method, the implicit midpoint rule, is employed. That is, the formula

$$\underline{y}_{n+1} = \underline{y}_n + h\underline{f}(x_{n+\frac{1}{2}}, (\underline{y}_n + \underline{y}_{n+1})/2) \tag{2.8}$$

is used. For linear problems with constant coefficients, $\underline{y}' = A\underline{y} + \underline{b}$, the formulae (2.1) and (2.8) yield identical results, and, indeed, a power series (2.4) for the error exists for the implicit midpoint rule. However the differential equations corresponding to (2.5) for the functions $\underline{c}_i(x)$ in (2.4) are slightly different in the case of (2.8). It is well-known that the trapezium rule is A-stable and the implicit midpoint rule has the same property, making it suitable for solving stiff equations. Lindberg (1972) gives two reasons for preferring the implicit midpoint rule. He presumes that the values $\underline{y}_{n+1}$ in formulae (2.1) and (2.8) will be determined by an iterative process such as a Newton iteration, and observes that, after calculating $\underline{y}_{n+1}$, $\underline{f}(x_{n+1}, \underline{y}_{n+1})$ must be calculated for use at the next step in the trapezium rule, whereas there is no similar requirement for the midpoint rule. To illustrate the other reason for preferring the implicit midpoint rule, consider the model equation $y' = \lambda(x)y$ where $\lambda(x) < 0$, so that $|y(x)|$ decreases as x increases. Using the trapezium rule we obtain

$$y_{n+1} = \left[\frac{1 + \frac{h}{2}\lambda(x_n)}{1 - \frac{h}{2}\lambda(x_{n+1})}\right] y_n, \qquad (2.9)$$

and to ensure that $|y_{n+1}| \leqslant |y_n|$ we must require (Gourlay 1970)

$$h\left[\lambda(x_{n+1}) - \lambda(x_n)\right] \leqslant 4. \qquad (2.10)$$

When $\lambda(x_n) < \lambda(x_{n+1})$ this condition restricts the possible size of h. The implicit midpoint rule has no such restriction. However, some problems with the implicit midpoint rule were outlined in Chapter 11, and it is still not clear which method ought to be used. Dahlquist (1975) has proposed some new ideas for the stability analysis of a class of methods which includes the implicit midpoint rule.

Now consider the model problem

$$y' = qy, \quad y(x_0) = y_0 \qquad (2.11)$$

where q is a complex constant such that $\mathrm{Re}(q) < 0$. For either formula (2.1) or (2.8), we have

$$y_{n+1} = \left[\frac{1 + \frac{h}{2}q}{1 - \frac{h}{2}q}\right] y_n. \qquad (2.12)$$

We observe from (2.12) that, when $\mathrm{Re}(\frac{hq}{2}) \ll 0$, we have $y_n \sim (-1)^n y_0$, and instead of the expected damped solution for (2.11) we obtain an oscillatory solution. From chapter 9, we recall that $\mathrm{Re}(hq) \ll 0$ for those components of a stiff system which are rapidly damped. Hence we would prefer the corresponding components of the approximate solution to be damped. With this fact in mind, we propose smoothing the approximate solution and hence, hopefully, damping the appropriate components. Lindberg (1971a) proposed a simple smoothing procedure namely replacing the approximate solution $\underline{y}_n$ by

$$\hat{\underline{y}}_n = (\underline{y}_{n-1} + 2\underline{y}_n + \underline{y}_{n+1})/4, \tag{2.13}$$

Note that this smoothing can be used in two entirely different ways. We could use (2.8) to compute $\underline{y}_0, \underline{y}_1, \ldots, \underline{y}_N$ then use the smoothing (2.13) passively to improve these values. Alternatively we could smooth after every $n(\geq 1)$ steps then use the smoothed value actively to compute later values of $\underline{y}_k$. In this active process, we must compute $\underline{y}_{n+1}$ twice, once before smoothing for use in (2.13), then again after smoothing in the continuing computation starting from $\hat{\underline{y}}_n$. From (2.12), for the test equation we have

$$y_n = \left[\frac{1 + hq/2}{1 - hq/2}\right]^n y_0$$

for all n, and hence

$$\begin{aligned}\hat{y}_n &= \frac{1}{4}\left[1 + 2\left(\frac{1 + hq/2}{1 - hq/2}\right) + \left(\frac{1 + hq/2}{1 - hq/2}\right)^2\right]\left[\frac{1 + hq/2}{1 - hq/2}\right]^{n-1} y_0 \\ &= \frac{1}{(1-hq/2)^2}\left[\frac{1 + hq/2}{1 - hq/2}\right]^{n-1} y_0 \\ &= \frac{1}{(1-(hq/2)^2)}\, y_n. \end{aligned} \tag{2.14}$$

Similarly if we perform M smoothings actively between y_0 and y_N we obtain

$$y_N(\text{smoothed}) = 1/(1-(hq/2)^2)^M\, y_N(\text{unsmoothed}), \tag{2.15}$$

and we observe that if $\text{Re}(hq/2) \ll 0$, the smoothed value of y_N can be heavily damped by a sufficient number of applications of the formula (2.13). However for nonlinear equations (2.2) and for $|hq/2| < 1$, we cannot draw such simple conclusions. For small values q, we are interested only in accuracy and smoothing may be counterproductive. Lindberg (1971b) considers the effect of using formula (2.13) (and a more accurate five-point smoothing formula). By considering inhomogeneous model equations, he shows that in an active smoothing process the formula (2.13) damps out the oscillating error but introduces a new oscillation of the order of the global discretization error. In the algorithm IMPEX2 given by Lindberg (1973) just one passive smoothing with formula (2.13) is used.

In the same way that the type and number of smoothings to be used must be decided when constructing an algorithm, so must the type and number of extrapolations to be used. Recall that in the Gragg-Bulirsch-Stoer method, essentially sufficient extrapolations are performed to achieve a required local accuracy and the extrapolated values are used in the continuing integration. If such a technique were used here we could encounter some stability problems (Stetter 1973, pp. 375-378). Instead, in the IMPEX algorithms passive extrapolation is used, and to ease the problem of estimating the error in the computed values only one extrapolation is employed and hence the method has $O(h^4)$ global discretization error. The stepsize used is controlled by an estimate of the local error in the extrapolated values but those values are used only as the computed output values, not in the continuing integration.

Finally in this section, we describe some of the computational details of the IMPEX2 algorithm (Lindberg 1973). The implicit midpoint rule is used to calculate two sequences of values $\underline{y}(x_k, h)$ and $\underline{y}(x_k, h/2)$, $k = 1, 2, \ldots$, with step h and h/2 respectively. The values $\underline{y}(x_{k-1} + h/2, h/2)$, $\underline{y}(x_k, h/2)$ and $\underline{y}(x_k, h)$ are obtained in turn for each value k using a modified Newton method to solve equation (2.8). That is, at each "step" starting with $\underline{y}(x_{k-1}, h)$ we perform one step of length h to obtain $\underline{y}(x_k, h)$, and starting with $\underline{y}(x_{k-1}, h/2)$ we perform two steps of length h/2 to obtain $\underline{y}(x_k, h/2)$. As usual an attempt is made to use the same Jacobian matrix $\partial\underline{f}/\partial\underline{y}$ for as many steps as possible in the Newton method. Starting values for the Newton iteration at each step are obtained by accurate extrapolation from computed values at previous steps, and within the current step as they become available. The two sequences of values are both passively smoothed using formula (2.13), and then used in a passive extrapolation employing formula (2.6). Accurate estimates of the function $\underline{c}_1(x)$ in the expansion (2.4) for the implicit midpoint rule are calculated from this information at each step and a backward difference representation of the solution $\underline{y}(x, h/2)$ and of $\underline{c}_1(x)$ is stored. Lindberg (1972a) describes an updating technique for these differences used in a step-by-step algorithm. The use of this information in an error estimate and control algorithm are given in Lindberg (1972b) and the use of the estimates in changing stepsize is described in Dahlquist and Lindberg (1973). We remark that most of the programming involved in this algorithm is conventional and similar to that described for many other methods earlier, but that the technique for changing stepsize is original. Dahlquist and Lindberg (1973) observe that to change stepsize at x_k one must interpolate to previously computed values. Alternative approaches are to interpolate on either unsmoothed or smoothed, and unextrapolated or extrapolated values. However all these approaches could be unsuccessful as it is crucial that the solution obtained with the new stepsize H should follow the path of a smooth solution through the previous values, and this is not guaranteed for these interpolated values. The basic idea used in IMPEX2 is to extrapolate from $\underline{y}(x_k, h)$ and $\underline{y}(x_k, h/2)$ to obtain not the values obtained using (2.6) (which is extrapolation to $\underline{y}(x_k, 0)$), but to obtain the values $\underline{y}(x_k, H)$ and $\underline{y}(x_k, H/2)$. Full details are given in Dahlquist and Lindberg (1973); basically, for any H we replace (2.6) by

$$\underline{y}(x_k, H) = \underline{y}(x_k, h/2) + (4(H/h)^2 - 1)(\underline{y}(x_k, h) - \underline{y}(x_k, h/2))/3, \quad (2.16)$$

and Dahlquist and Lindberg show that this extrapolation gives values on a smooth solution.

3. A COMPARISON OF METHODS FOR STIFF PROBLEMS

In this section, we briefly compare algorithms for stiff systems of ordinary differential equations. Since the appearance of the report by Hull, Enright, Fellen and Sedgwick (1972) which compares algorithms for non-stiff problems and draws some quite firm conclusions, there have been attempts to do the same for algorithms for stiff equations. We discuss some of these attempts later, but we observe that the state of the art in algorithms for stiff problems is not the same as for non-stiff problems and it is probably too soon to draw any firm conclusions. First, we discuss the qualitative structure of some typical problems as it seems to us that it is impossible to discuss the usefulness of various algorithms without an appreciation of the types of problems involved. In the next two chapters some practical problems are discussed, and a much larger number are described in Bjurel, Dahlquist, Lindberg, Linde, and Oden (1970).

Maybe the first and one of the most important questions about a stiff problem is its degree of nonlinearity. In general, to solve linear problems we would not require the sophisticated methods which have been described previously. Often, a linear problem will arise at an intermediate stage in the solution of a partial differential equation. It is likely that the (possibly time-dependent) Jacobian of the system will be large and have a special banded structure and then the best recommendation will usually be a method specially constructed for such a problem. There is a discussion of appropriate methods in chapter 14.

Many of the problems which have been used for comparison purposes involve small but nonlinear systems of differential equations. Enright, Hull and Lindberg (1974) use several such problems and they note that the form of the nonlinear coupling of the components of the solution can affect the behaviour of an algorithm. For example, they observe that when the components of a system can be divided into two classes, one transient and the other smooth, and when the differential equations for the transient components contain nonneglible nonlinear terms in the smooth components, then an algorithm of the extrapolation type described earlier in this chapter behaves rather poorly. This was essentially predicted in Dahlquist and Lindberg (1973). Additional entirely different problems arise with large nonlinear stiff systems. Usually the Jacobian matrix of such a system is very sparse and any useful algorithm must take cognizance of this fact. An algorithm has been described in Gear (1971c) and many others developed by workers in the fields of application where such problems arise. The production of good all-purpose algorithms for large nonlinear stiff problems should be a major area of future research. In nonlinear problems we must use an iterative method of solution at each step, for example some variant of Newton's method. In the Newton method we need an estimate of the Jacobian matrix and this matrix must be updated occasionally. If a large part of the Jacobian is constant, or essentially constant, as will often be the case after the transient phase, then this fact can be exploited to reduce the matrix algebra involved (Robertson 1975a).

In some problems it is feasible to write the differential equation (2.2) in the form

$$\underline{y}' = J\underline{y} + \underline{g}(x, \underline{y}) \tag{3.1}$$

where J is (at least locally) a constant matrix which contains all the information about the stiffness of the problem and $\underline{g}$ is a small nonlinear term. Algorithms have been devised which attempt to use knowledge of J so as to transform the problem to a non-stiff problem. Algorithms of this type are the extrapolation method of Odén (1971) and the Runge-Kutta method of Lawson and Ehle (1972). The latter method does not seem very promising for nonlinear problems (Enright et al 1974).

Another source of possible difficulty for some algorithms is that the Jacobian matrix $\partial \underline{f}/\partial \underline{y}$ have eigenvalues λ_i with significant imaginary parts. If, for some eigenvalue λ_i, $\arg(\lambda_i) \sim \pi/2$, this eigenvalue may contribute a significant oscillatory component to the solution and the algorithm must be capable of representing this component adequately and stably. This aim can only be achieved by taking a small enough step-length with any method. If there is no oscillatory component in the solution, then only stability matters and a method which is not A-stable could perform poorly. Enright et al (1974) show that their version of Gear's method does not deal adequately with such problems. This is not surprising since the high-order formulae used in Gear's method have poor stability properties near the imaginary axis.

It is sometimes possible to exploit knowledge of the location of the eigenvalues of the Jacobian. For example, if the eigenvalues can be grouped into two sets each containing eigenvalues of approximately the same magnitude, but the magnitudes of the eigenvalues in the two sets differ greatly, then we might use a method designed for just this case. Two such methods are the extrapolation method of Odén (1971) and the exponentially fitted methods of Liniger and Willoughby (1970).

Finally, we describe some of the comparative reports on algorithms for stiff systems. First, we describe a result due to Lindberg (1974a). He considers the problem

$$\underline{y}' = 10^4\begin{pmatrix} 1-2e^{-t} & te^{-t} \\ -te^{-t} & 1-2e^{-t} \end{pmatrix}\underline{y}, \quad \underline{y}(0) = \begin{pmatrix} 1 \\ 1 \end{pmatrix} \tag{3.2}$$

and shows that $\|\underline{y}\|_2$ decreases from 1.4 at $t = 0$ to 4.10^{-1333} at $t = 0.7$ and then increases to 10^{34744} at $t = 10$. In contrast a numerical solution computed using Gear's method with local error tolerance 10^{-6} initially decreases to 10^{-10} and then changes little. He remarks that this would happen with any method when eigenvalues of the Jacobian change sign after the transient stage so that the step being used at this point is relatively large. In fact he demonstrates why the same behaviour would occur for the backward Euler method and the trapezium rule. This type of problem has not been included in the tests described below.

There have been two major attempts to compare methods for stiff problems in the way in which methods for non-stiff problems were compared by Hull et al (1972). They are described in the reports by Ehle (1972) and Enright et al (1974). Ehle (1972) compares several implicit Runge-Kutta methods with explicit Runge-Kutta methods using an explicit estimate of the Jacobian matrix and a version of Gear's method. In contrast, Enright et al (1974) compare an implicit Runge-Kutta method, an explicit Runge-Kutta method using the Jacobian explicitly, a version of Gear's method, Enright's second derivative method and an extrapolation method based on the trapezium rule. In both reports all the methods are compared over a large number of different types of problems and, indeed, there is some overlap in the problems chosen. The criteria for comparison in the two reports are similar and include, for example, the number of calls to the subroutine for evaluating $\underline{f}(x, \underline{y})$ and the number of matrix inversions required. Enright et al (1974) use an averaged error per unit step criterion for stepsize selection. There is some evidence (Lindberg 1974b) that error per unit step techniques can lead to the selection of far from optimal stepsize sequences. As Enright et al (1974) say, this possibly makes their results rather less realistic than they might otherwise be, but they have no evidence that this has happened. These reports do not come down in favour of any single method but seem to imply that the Runge-Kutta methods are not competitive overall.

Lindberg (1973) compares his IMPEX2 algorithm with some of the algorithms in Ehle's report, again without drawing any conclusions. These comparisons use many different versions of Gear's method. Craigie (1975) discusses a divided difference version of Gear's method with theoretically better stability properties than the original scaled derivative Gear program (Gear, 1971b). Craigie compares his divided difference algorithm with Gear's DIFSUB program on a selection of problems from Enright et al (1974) but he finds no great difference in performance.

13

NUMERICAL INTEGRATION OF SYSTEMS OF STIFF ORDINARY DIFFERENTIAL EQUATIONS WITH SPECIAL STRUCTURE

1. INTRODUCTION

In this chapter we are concerned with the numerical integration of systems of first order ordinary differential equations in which some of the variables have response times which are very short on the time scale of the interesting behaviour of the system. It is characteristic of such systems that these fast responding variables are "stable", in the sense that if the slower variables were at any time clamped (i.e. held fixed), the fast responding variables would quickly tend toward and remain at equilibrium values.

Systems of equations having these properties arise in a variety of applications, including chemical reaction kinetics, particularly where very reactive intermediate species are produced, guidance and control problems, electrical transmission networks, and heat and matter transfer. Systems with fast and slow motions are by definition stiff and are frequently studied by computational methods. It was shown in Chapter 9 that stability requirements dictate the use of an implicit method to achieve efficiency by enabling the step length to be progressively increased as the solution proceeds (see for example Robertson (1966), Dahlquist (1969) and Prothero and Robinson (1974)). Implicit methods involve the solution of the corresponding set of non-linear "algebraic" equations at each step of the integration and these are commonly solved by a modified Newton-iteration procedure or the fixed chord method where the iteration matrix is based on an approximation $f' + B$ to the Jacobian f' in a neighbourhood of the solution. Robertson and Williams (1975) have drawn attention to the sensitivity of predictions made with the use of calculated derivatives and analysed the factors affecting local convergence rates with special emphasis on the role of the perturbation B. We showed broadly that the smaller B is, the faster is the rate of convergence. The normal experience is that an iteration matrix gives satisfactory convergence over a number of integration steps before slow convergence forces a re-evaluation of the Jacobian. If the system is large, the associated inversions or LU splits required to provide solutions of the iteration equations represent a large proportion of the computations. In this chapter we show how some of this work may be avoided when the Jacobian has a special structure which enables satisfactory convergence to be re-established by updating a partition of the Jacobian. We are also able to show that the perturbation B may have a somewhat similar structure to that of the Jacobian. This enables us to establish a tighter, readily computable approximation giving a sufficient condition for convergence and a better computable estimate of the rate of convergence, as well as giving an indication of a suitable partition for an updating procedure.

2. SPECIAL STRUCTURE

The structure and general behaviour of our problem is closely related to that of singular perturbations of the initial value problem. For the purposes of preliminary orientation we consider the simplest type of such problems consisting of a pair of scalar equations of the form

$$\frac{dx}{dt} = f(x,y),$$

$$\varepsilon \frac{dy}{dt} = g(x,y),$$

where ε is small and positive and where the initial conditions are given. For a qualitative description of the behaviour of this system we refer to Wasow (1965). Suffice to say that in appropriate circumstances the solution rapidly approaches an ε-neighbourhood of the curve $g(x,y) = 0$. We identify the x and y variables with the slow and fast motions respectively and observe that the Jacobian of the system,

$$\begin{pmatrix} f_x & f_y \\ \varepsilon^{-1} g_x & \varepsilon^{-1} g_x \end{pmatrix},$$

has in general, much larger elements in the bottom row. This structure of the Jacobian is typical of the type we wish to consider. We are not however concerned with the analytical properties of the system for limiting values of the parameters. We note that this matrix can be transformed by scaling of rows and columns to one with a much larger second column. A linear transformation of the dependent variables of the differential equation produces a new system of equations whose Jacobian is similar to the original and with effects which may correspond to scaling operations on the rows and columns. In fact, we prefer to take the column form of structure.

For our purposes, a more appropriate description of the systems of interest is in terms of the Jacobian and higher derivatives since we are concerned essentially with local properties in the neighbourhood of the solution in the asymptotic region where the fast transients have died away. Our problem may therefore be written as the numerical integration of the equations

$$\frac{dy}{dx} = f(y), \quad y(0) = y_0 , \quad y \varepsilon R^n \tag{2.1}$$

where the Jacobian f' has eigenvalues λ_i, $i = 1,2,\ldots,n$ such that $Re(\lambda_i) < 0$ and $\max\limits_{i,j} \dfrac{Re(\lambda_i)}{Re(\lambda_j)}$ is large. We suppose further that y is partitioned into

$$\begin{pmatrix} y_1 \\ y_2 \end{pmatrix}, \quad y_1 \varepsilon R^{n_1}, \quad y_2 \varepsilon R^{n_2}, \ (n_1+n_2= n),$$

such that the conformal partition of f' by columns into

$$f' = \begin{pmatrix} f_{11} & f_{12} \\ & \\ f_{21} & f_{22} \end{pmatrix} = (f'_1, f'_2),$$

satisfies

$$||(f'_1, 0)|| \ll ||(0, f'_2)||. \tag{2.2}$$

Further conditions on f_{22} ensure the stability of the fast motions associated with y_2. When f' varies slowly there is little need to adjust its value for computational purposes but when the changes in f' along the solution affect convergence properties and when the structure described by (2.2) is maintained along the solution there is the possibility of increased efficiency by confining updating to the partition f'_2. More precisely (2.2) will remain true over a range when $\frac{df'}{dx}$ or f"f' has the same structure as f'.

The computational problem of solving such systems, particularly in the asymptotic region, is essentially that of the efficient solution of systems of non-linear algebraic equations since the integration formula is of implicit type. We therefore consider the solution of equations of the form

$$F(y) = y - h\gamma f(y) - \psi, \quad \gamma > 0, \tag{2.3}$$

which result from application of an implicit numerical integration formula to equation (2.1). h is the step length, γ is a constant associated with the particular formula and similarly ψ depends on a combination of past values of the function and derivative. A practical method of solving the equations is to use the parallel chord method which may be written

$$y^{k+1} = y^k - M^{-1}F(y^k), \quad k = 0,1,2,\ldots, \tag{2.4}$$

where y^0 is a given initial approximation and where M is a non-singular constant matrix which may be taken to be an approximation to the derivative at the solution point y*. Thus

$$M = F'(y^*) + E, \tag{2.5}$$

for some perturbation E, which in practice is usually given by

$$E = M - F'(y^*) = F'(y^\dagger) - F'(y^*), \tag{2.6}$$

where $y^\dagger$ is some previously determined point on the solution.

During the period of very fast transients it may be possible to employ an explicit formula $(\gamma = 0)$, in which case no question of equation-solving arises; but as the transients decay, the interval of integration can generally be increased to the point where stability considerations of the approximation become important and an implicit formula such as (2.3) has to be employed. For small values of step length h this can be satisfactorily solved by the method of repeated substitutions corresponding to the use of $M = I$ in (2.4). As the interval is further increased, the convergence properties of this method deteriorate but these may be restored if $F'(y)$ is diagonally dominated by taking M to be the diagonal of $I - h\gamma f'(y) + E$, but in most cases, and as the step length is still further increased, it is necessary to take account of the off-diagonal elements of f' and at this stage we may take advantage of structure.

3. A PARTITIONED APPROXIMATE INVERSE

The matrix M^{-1} arising in the iterative method (2.4) is the focus of our attention and, as we have mentioned, it frequently arises by inversion or appropriate LU split of the Jacobian $F'(y^\dagger)$ at some point $y^\dagger$ of the approximate solution. Given a partition $\begin{pmatrix} y_1 \\ y_2 \end{pmatrix}$ of the dependent variables y we take conformal partitions of the Jacobian F' and write

$$F' = \begin{pmatrix} F_{11} & F_{12} \\ F_{21} & F_{22} \end{pmatrix} = \begin{pmatrix} a & b \\ c & d \end{pmatrix},$$

and formally write the inverse as

$$(F')^{-1} = \begin{pmatrix} (a - bd^{-1}c)^{-1} & -(a - bd^{-1}c)^{-1}bd^{-1} \\ -(d - ca^{-1}b)^{-1}ca^{-1} & (d - ca^{-1}b)^{-1} \end{pmatrix},$$

$$= \begin{pmatrix} \alpha^{-1} & -\alpha^{-1}bd^{-1} \\ -\delta^{-1}ca^{-1} & \delta^{-1} \end{pmatrix}. \qquad (3.2).$$

Recall that the structure introduced in (2.2) requires that b and d are much "larger" than a and c and that b and d are more sensitive to changes along the solution. This motivates considerations of the iteration matrix

$$M = \begin{pmatrix} I & b \\ c & d \end{pmatrix}. \qquad (3.3).$$

Formally, we have

$$M^{-1} = \begin{pmatrix} I + b\delta_I^{-1}c & -b\delta_I^{-1} \\ -\delta_I^{-1}c & \delta_I^{-1} \end{pmatrix} \qquad (3.4)$$

where $\delta_I = d - cb$.

The corresponding iteration in terms of the partition of y and F becomes

$$\begin{aligned} y_1^{k+1} &= y_1^k - F_1(y^k) - b(y_2^{k+1} - y_2^k) \\ y_2^{k+1} &= y_2^k + \delta_I^{-1}(cF_1(y^k) - F_2(y^k)) \end{aligned}, \qquad (3.5)$$

or

$$y_1^{k+1} = h\gamma f_1(y^k) + \psi_1 - b(y_2^{k+1} - y_2^k)$$

$$, \qquad (3.6)$$

$$y_2^{k+1} = y_2^k + \delta_I^{-1}(cF_1(y^k) - F_2(y^k))$$

in terms of the integration equations (2.3). This iteration has a Newton-like character for fast motions and resembles repeated substitution with correction terms for the slow motions. This is a very plausible algorithm when we recall that the slow motions have much longer time constants so that repeated substitution of $F_1 = 0$ is rapidly convergent whereas the fast motions, corresponding to $F_2 = 0$, require local derivative information to be incorporated in the iteration for rapid convergence. An example will be presented later which shows that this iteration can successfully be employed. If the dimensions of δ_I are much less than those of the full matrix M ($n_2 << n_1$) considerable computational savings accrue. This is however only one stage in the integration, for, as the step length h is further increased, it will become expedient to take account of the structure of the partition a. If a remains diagonally dominated we may take advantage of this. If $a = a_d + e_a$ where a_d is a diagonal matrix the corresponding modifications to (3.4) may provide adequate convergence at the expense of a minimum of additional computation. Again, we may consider the partition y_1 to be further partitioned into fast and slow motions when an approximation of the form (3.4) will be appropriate for a^{-1}. In this way we can introduce a spectrum of motions into the problem.

When a full inverse is employed, the Sherman-Morrison-Woodbury formula, (c.f. Ortega and Rheinboldt 1970), may be used to update this for changes in a partition of the derivative. In a later section on convergence we give results which provide comparison criteria for the effectiveness of such updating.

4. CHEMICAL REACTION KINETIC EQUATIONS

Chemical reactions are conventionally symbolised by the stoichiometric equations which give the proportions of the various chemical species participating in the reactions. Thus, if Y_i is a chemical species, a reaction is symbolised by the "equation"

$$\sum_i \alpha_i Y_i = 0 \tag{4.1}$$

where conventionally, $\alpha_i \gtrless 0$ for a product or reactant respectively, α is the stoichiometric vector of the reaction which we further decompose to $\alpha = \pi - \rho$ where π and ρ are positive vectors referring to products of reaction and reactants respectively and where we assume $\pi_i \rho_i = 0, \forall i$ (a species is not both product and reactant in the same reaction: catalysts are not considered). In the general case, we are concerned with a number of reactions which can be accommodated by a matrix α whose columns are the stoichiometric vectors for the individual reactions. If the jth reaction is allowed to proceed to an extent ξ_j, the concentrations of the chemical species are then given by

$$y = y_0 + \alpha\xi, \tag{4.2}$$

where lower case letters denote the concentrations of corresponding species and where y_0 is the initial concentration. Differentiating w.r.t. time we obtain the rate equations in the form

$$\dot{y} = \alpha\dot{\xi} \quad . \tag{4.3}$$

$\dot{\xi}$ is the vector of reaction rates r. For elementary reactions the jth component of r, corresponding to the jth reaction, is a product of concentration terms over the species involved as reactants and raised to the power of the corresponding coefficient of the stoichiometric vector. A constant term is also incorporated so that

$$r_j = k_j \prod_i y_i^{\rho_{ij}} \quad . \tag{4.4}$$

This rate expression corresponds to the law of mass action and the constant k is the reaction kinetic constant (we are here treating the case of isothermal reaction). For a further description see Aris (1969) and for a comprehensive treatment we refer to the series "Comprehensive Chemical Kinetics", Ed Bamford and Tipper (1969).

5. PROPERTIES OF REACTION KINETIC SYSTEMS

We have made no mention of the dimensions of the vector and matrices involved in (4.2) for example, but a great variety of situations occur in practice. The essential requirement for a unique physical situation is that the column rank of α is a maximum or equivalently that the reactions form an independent set. An interesting situation is the case where α is a square matrix but is not of full rank. Suppose for example that $\beta^T \alpha = 0$ where β^T is an appropriate matrix. Premultiplication of (4.3) by β^T shows that the system has a number of linear invariants. These may have direct physical interpretations in terms of conservation of fragments of species and groups of species. These relations may be used to reduce the order of the systems in terms of concentrations.

Alternatively, or when this has been done, so that α is square and of full rank, there are clearly two sets of differential equations which can be solved to describe the behaviour of the system. (4.3) and (4.4) can be solved for the concentration vector y.

Alternatively,

$$\dot{\xi} = r \tag{5.1}$$

with the concentration variables explicit, as in (4.4), for r_j obtained from (4.2). Apart from the advantages of explicit formulation in terms of reactions, (5.1) has the advantage that the ξ variables are monotonically increasing and can be easily bounded whereas the concentration variables y, although positive, are not so simply analysed.

Theoretical reaction kinetic systems have a number of interesting properties some of which are discussed by Wei & Prater (1962) and Krambeck (1970). Of particular interest for numerical analysis applications is the question of asymptotic stability. The Liapounov function gives a method for determination of stability characteristics. In the context of Statistical Mechanics the Gibb's Free Energy and Entropy functions provide examples of Liapounov functions applicable to particular thermodynamic situations. For a fuller discussion of this aspect we refer to Wei (1965).

6. FAST REACTIONS ASSOCIATED WITH VERY REACTIVE SPECIES

We now assume a partition of the concentration vector y into $\begin{pmatrix} y_1 \\ y_2 \end{pmatrix}$ such that the components of y_2 are concentrations of very reactive species. Other things being equal, we expect $||y_2|| << ||y_1||$ on the physical basis that very reactive species quickly react with something and consequently their concentrations never remain very large. There is a corresponding partition of the rate vector r into fast and slow reactions $\begin{pmatrix} r_1 \\ r_2 \end{pmatrix}$. r_2 is a vector of rates in which very reactive species participate as reactants. The components of r_2 have rate constants k which are correspondingly large. We now turn to the Jacobian of the derivatives of the concentration equations (4.3). The Jacobian may be written

$$J = \alpha r' \tag{6.1}$$

when r' is the derivative of r with elements

$$\frac{\partial r_i}{\partial y_j} = \frac{\rho_{ji} r_i}{y_j} \tag{6.2}$$

If y_j is a very reactive species y_j^{-1} is large and this element of the Jacobian is relatively enhanced. Using the conformal partitions of J we have,

$$J = \begin{pmatrix} \alpha_{11}\frac{\partial r_1}{\partial y_1} + \alpha_{12}\frac{\partial r_2}{\partial y_1} & \alpha_{12}\frac{\partial r_2}{\partial y_2} \\ \\ \alpha_{21}\frac{\partial r_1}{\partial y_1} + \alpha_{22}\frac{\partial r_2}{\partial y_1} & \alpha_{22}\frac{\partial r_2}{\partial y_2} \end{pmatrix} . \tag{6.3}$$

By definition, $\frac{\partial r_1}{\partial y_2} = 0$, since a slow reaction can not have a reactive species as a reactant. (6.3) "shows" that the columns associated with differentiation with respect to y_2 are much "larger" than the remainder. The system is sensitive to the very reactive species y_2. The Jacobian of such systems thus has the special structure discussed in §2 and there is the opportunity to apply the economical iteration (3.6). Furthermore, the changes in the Jacobian are mainly associated with the fast reactions, so that adequate convergence may be maintained by updating the inverse mainly with respect to the corresponding columns.

It is clear that chemical reaction kinetic systems with fast and slow reactions can be expressed in terms of a small parameter ε (or large reaction kinetic constant k) which gives the problems the structure of singular perturbations. Associated with the full problem set out in §2 there is a reduced problem obtained by setting $\varepsilon=0$. This approximation is known in some contexts in physics and chemistry as the "Steady State Approximation". The fast variables rapidly establish equilibrium with the slower motions, and so their motion is well described by the algebraic relation $g(x,y) = 0$. Stockmayer (1944) discusses an application in polymerisation kinetics. Levin and Levinson (1956) discuss the mathematical conditions under which this approximation is valid. Hoppensteadt (1971) derives further properties of such systems and develops solution techniques of boundary layer type. Cole (1968) discusses applications. The approximation is not always correctly used and Habets (1974) has discussed a consistency theory of singular perturbations. Edsberg (1974) describes some applications and makes comparisons with other methods. Lapidus, Aiken and Liu (1974) also discuss the occurrence and numerical solution of systems with widely varying time constants. Expansion methods for this problem were devised by Vasil'eva (1963) and others and extended by O'Malley (1973). Dahlquist (1968) gives a numerical method for this problem and Miranker (1973) uses numerical methods of boundary layer type and develops these to cope with parameter-less systems.

7. APPLICATIONS

We now give a number of examples of systems of stiff differential equations which exhibit the structure defined in (2.2) of a Jacobian which has column partitions of widely separated "magnitudes". For further details of some of the examples discussed here and a comparative review of numerical methods for the solution of stiff systems see Enright, Hull & Lindberg (1974).

Example 1.

Gear (1968) discusses the application of stiffly-stable integration formulae based on backward difference approximation of the derivative and uses the following example (with a sign change for y_3) to illustrate the method.

$$\frac{dy}{dx} = f(y), \qquad y \in \mathbf{R}^3,$$

where

$$f_1 = -.013y_1 + 10^3 y_1 y_3,$$

$$f_2 = +2.5 \times 10^3 y_2 y_3 ,$$

$$f_3 = .013y_1 - 10^3 y_1 y_3 - 2.5 \times 10^3 y_2 y_3 ,$$

with initial conditions

$$y(0) = \begin{pmatrix} 1 \\ 1 \\ 0 \end{pmatrix}.$$

This application is from chemical reaction kinetics and y_3 represents the concentration of a very reactive species which is an intermediate in the course of the reaction and always stays small. y_1 and y_2 are monotonically decreasing and increasing respectively. y_3 increases to a maximum and thereafter is monotonically decreasing. It is not difficult to show that $y_3 < 1.3 \times 10^{-5}$. The Jacobian of the time derivative f is given by

$$J = \begin{pmatrix} -.013 + 10^3 y_3 & 0 & 10^3 y_1 \\ 0 & 2.5 \times 10^3 y_3 & 2.5 \times 10^3 y_2 \\ .013 - 10^3 y_3 & -2.5 \times 10^3 y_3 & -10^3 y_1 - 2.5 \times 10^3 y_2 \end{pmatrix}.$$

Due to the smallness of y_3, the last column of this matrix has elements which are very much larger than the remainder. y_1 and y_2 are clearly associated with slow motion whilst y_3 is identified with the fast motion. In terms of the structure of (2.1), $n_1 = 2$ and $n_2 = 1$. The matrix updating procedure associated with the partitioned approximate inverse (3.4) no longer requires matrix inversions. This is most clearly seen from (3.6) where δ_I from being a matrix requiring inversion, has degenerated to a scalar. This example illustrates well the features of the Jacobian which are typical of chemical reaction kinetic systems with very reactive intermediates. Another common feature of reaction kinetic systems is the polynomial character of f. Here f is quadratic in the elements of y so that f" is a constant operator and $f'(y^\dagger)-f'(y^*)$ is linear in $(y^\dagger-y^*)$. It will readily be seen that this implies that the perturbation $E = F'(y^\dagger) - F'(y^*)$ has the same structure as the Jacobian so that the matrix M of the iteration and the corresponding inverse are updated mostly by changes in y_1 and y_2 and less frequently by a change in y_3.

Robertson (1966) discusses a similar system from reaction kinetics in three dimensions and also having quadratic derivatives. This system has an invariant but is otherwise similar to the above.

Example 2.

Lindberg (1974) considers a system of dimension 4, where the derivatives are given by

$$f_1 = -k_1y_1y_2 + k_3y_3,$$

$$f_2 = k_3y_3 - 2k_2y_2^2 - k_1y_1y_2,$$

$$f_3 = -k_3y_3 + k_1y_1y_2,$$

$$f_4 = -k_4y_4 + k_2y_2^2,$$

and where

$$k = \begin{pmatrix} 10^2 \\ 10^4 \\ 1 \\ 1 \end{pmatrix} \text{ and } y(0) = \begin{pmatrix} 1 \\ 1 \\ 0 \\ 0 \end{pmatrix}.$$

If we make the change of variable $\eta = Py$ where P is the permutation matrix interchanging 1 and 3 and 2 and 4, the Jacobian becomes

$$P^{-1}JP = \begin{pmatrix} -1 & 0 & k_1\eta_4 & k_1\eta_3 \\ 0 & -1 & 0 & 2k_2\eta_4 \\ 1 & 0 & -k_1\eta_4 & -k_1\eta_3 \\ 1 & 0 & -k_1\eta_4 & -k_1\eta_3-4k_2\eta_4 \end{pmatrix}.$$

We observe that η_3 and η_4 are associated with the fast motions, so that matrix updating may be confined to the last two columns. Eventually, η_4 becomes sufficiently small so that the last column alone requires frequent updating when the associated matrix inversion degenerates to scalar operations.

8. LOCAL CONVERGENCE OF THE PARALLEL CHORD ITERATION

The parallel chord iteration for solution of the equation $F(y) = 0$ may be written in the form

$$M(y^{k+1} - y^k) = -F(y^k), \quad k = 0,1,..., \tag{8.1}$$

where y^0 is given and where the matrix M may be taken to be an approximation to the derivative at y* so that

$$M = F'(y^*) + E, \tag{8.2}$$

for some perturbation E.

Sufficient conditions for the local convergence of the iteration $y^{k+1} = G(y^k)$ are given by Ostrowski's theorem (Ortega and Rheinboldt 1970). In particular, this includes the essentially necessary condition $\rho(G'(y^*)) < 1$, where ρ is the spectral radius and y* is a fixed point of G. In the following we show how this may be replaced by a more easily computable condition which sometimes implies a much stronger condition on the spectral radius of $G'(y^*)$. We also consider some modifications which can relax this and so maintain the benefit of an easily computable condition without being too strong for practical use. The proofs of the following theorems together with other details are given elsewhere (Robertson 1975)

<u>THEOREM 1</u> For the iteration (8.1) let $F:R^n \to R^n$ be differentiable at the solution point y* of the equations $F(y) = 0$. Assume that $F'(y^*)$ is non-singular and that the perturbation E satisfies

(i) $||F'(y^*)^{-1}|| \, . \, ||E|| < \frac{1}{2}$,

where $||\,.\,||$ is the matrix norm subordinate to a given norm on R^n, then the sequence of iterates $\{y^k\}$ is well defined and converges to y^* for all y^0 sufficiently close to y^*.

An estimate of the spectral radius of $G'(y^*)$ is

$$\rho(G'(y^*)) \leqslant ||M^{-1}E|| \leqslant \frac{||F'(y^*)^{-1}E||}{1 - ||F'(y^*)^{-1}E||} \leqslant \frac{||F'(y^*)^{-1}||\,.\,||E||}{1 - ||F'(y^*)^{-1}||\,.||E||} \tag{8.3}$$

<u>Corollary</u>. If P and Q are non-singular matrices and the conditions of the theorem apply to the iteration

$$QMP(z^{k+1} - z^k) = -QF(Pz^k), \quad k = 0,1,\ldots, \tag{8.4}$$

where $z^0 = P^{-1}y^0$ is given, then the iteration (8.1) is locally convergent

Remark: From the corollary it follows that condition (i) of the theorem may be replaced by the condition

$$\text{(ii)} \quad ||P^{-1}F'(y^*)^{-1}Q^{-1}||\,.\,||QEP|| < \tfrac{1}{2},$$

where P, Q are non-singular matrices.

The corresponding estimate of the spectral radius is

$$\rho(G'(y^*)) \leqslant \frac{||P^{-1}F'(y^*)^{-1}Q^{-1}||\,.\,||QEP||}{1 - ||P^{-1}F'(y^*)^{-1}Q^{-1}||.||QEP||}\;. \tag{8.5}$$

When the matrix M^{-1} of the parallel chord method is taken to be a perturbation of the inverse Jacobian so that

$$M^{-1} = F'(y^*)^{-1} + W \tag{8.6}$$

the following theorem applies.

<u>THEOREM 2</u> For the iteration (8.1) let $F:R^n \to R^n$ be differentiable at the solution point y^* of the equations $F(y) = 0$. Assume that $F'(y^*)$ is non-singular and that M is non-singular with W defined as in (8.6) satisfying

$$\text{(iii)} \quad ||F'(y^*)||\,.\,||W|| < 1,$$

then the sequence of iterates $\{y^k\}$ is well defined and converges to y^*

for all y^0 sufficiently close to y^*.
As estimate of the spectral radius of $G'(y^*)$ is given by

$$\rho(G'(y^*)) \leqslant ||F'(y^*)|| \,.\, ||W|| \,. \tag{8.7}$$

Corollary. Let P and Q be non-singular matrices and let the conditions of the theorem apply to the iteration (8.4) where $z^0 = P^{-1}y^0$ is given, then the iteration (8.1) is locally convergent.

It follows that condition (iii) of Theorem 2 can be replaced by

$$\text{(iv)} \quad ||QF'(y^*)P|| \,.\, ||P^{-1}WQ^{-1}|| < 1 \,.$$

The corresponding estimate of the spectral radius is

$$\rho(G'(y^*)) \leqslant ||QF'(y^*)P|| \,.\, ||P^{-1}WQ^{-1}|| \tag{8.8}$$

The two theorems are related but independent. It is not difficult to show that when the conditions of Theorem 1 apply, $||F'(y^*)W|| < 1$, but the particular case of E a scalar multiple of $F'(y^*)$ shows that (iii) is not generally valid. Similarly the conditions of Theorem 2 do not imply (i).

Finally we remark that the practical usefulness of the transformations introduced in (8.4) is in obtaining a readily computable and accurate estimate of the spectral radius $\rho(G'(y^*))$. Also, by considering partitions of E which are set to zero it enables us to determine accurately the convergence effects of updating the matrix M in a variety of ways.

9. EQUILIBRATION AND CONDITION OF F'

Stiff differential equations have a Jacobian F' which has a poor condition number $\kappa(F') = ||F'|| \,.\, ||F'^{-1}||$ for sufficiently large step-size h. A linear transformation of the dependent variables y gives a new system of equations with a Jacobian which is similar to F' and therefore has the same spectrum. The matrix M of the equation (8.1) defining the iteration is badly equilibrated (Wilkinson 1965, p. 192), and this can be improved by scaling of rows and columns

(Bauer 1963). Such scaling operations can be accommodated by the transformations we have introduced in the equivalent iteration (8.4). The estimate (8.3) for the spectral radius of $G'(y^*)$ may be written

$$\rho(G'(y^*)) \leqslant \frac{\kappa(F'(y^*)) \quad ||E|| \;/\; ||F'(y^*)||}{1 - \kappa(F'(y^*))||E|| \;/\; ||F'(y^*)||} \; . \tag{9.1}$$

The term $|| \; E \; || \; / \; ||F'(y^*)||$ is interpreted as the relative perturbation. When the condition number of the Jacobian can be reduced by equilibration of rows and columns without adversely affecting the relative perturbation there is the opportunity of an improved estimate for $\rho(G'(y^*))$. The corresponding estimate in terms of W the perturbation of the inverse may be written

$$\rho(G'(y^*)) \leqslant \kappa(F'(y^*)) \quad ||W|| \;/\; ||F'(y^*)^{-1}|| \; . \tag{9.2}$$

$||W \; || \; / \; ||F'(y^*)^{-1}||$ is interpreted as the relative perturbation of the inverse and a similar remark applies to the improvement of this estimate by use of transformation matrices P and Q.

10. NUMERICAL RESULTS

A system of stiff differential equations has been solved numerically by a standard programme developed at ICI Ltd. using an algorithm of Robertson (1967). Briefly, the trapezoidal rule is used to integrate the equations and the parallel chord method is used to solve the non-linear algebraic equations at each step. Prediction is carried out by extrapolation from function values as described by Robertson and Williams (1975) and automatic error control and step adjustment is achieved by the Milne-type error test. A maximum number of four iterations is allowed to attain a given accuracy as measured by the difference between successive iterates. When convergence is unsatisfactory the inverse is updated by evaluation of the Jacobian at the last accepted point. If this is still unsatisfactory the step length is halved. When convergence has been achieved, truncation error estimates are made and further adjustments to the step-length are considered. It is necessary to limit the frequency with which the step-length can be adjusted upwards in order to prevent excessive oscillations in step-length-upwards for reasons of accuracy and downwards due to slow convergence. A general measure of the amount of work involved in solving the non-linear equations over a range is the number of inversions and the number of iterations. These parameters are affected by the number of changes in step-length, since a new inverse is determined whenever the step-length is altered. However, in any comparison of inverse updating procedures we expect the genuine adjustments of step-length due to accuracy considerations to remain a common background since this is a function of the behaviour of the differential equation. We can therefore, with some allowance, attribute differences in these parameters to the different updating procedures. As mentioned previously the efficiency of any updating procedure depends on the structure and varies with the step-length. For small enough values of the step-length the iteration matrix M of (2.4) is diagonal dominant and as the step-length is increased the full effects of structure are felt. In this illustrative computation we have confined ourselves to a simple strategy with a view to demonstrating the potential efficiency to be gained from use of inverse updating allowing for structure.

Consider the time behaviour of the six chemical species Y_1, Y_2, Y_3, Y_4, P_1, and P_2 which participate in four chemical reactions which are described in the conventional way by the "equations",

$$\begin{aligned}
&Y_1 &&\overset{k_1}{\rightarrow} &&Y_2 + Y_3 \;, \\
&Y_1 + Y_3 &&\overset{k_2}{\rightarrow} &&Y_4 \;, \\
&Y_4 &&\overset{k_3}{\rightarrow} &&Y_3 + P_1 \;, \\
&Y_2 + Y_3 &&\overset{k_4}{\rightarrow} &&P_2 \;.
\end{aligned} \tag{10.1}$$

Denoting the concentrations of the chemical species by lower case letters we obtain the system of differential equations

$$\frac{dy}{dt} = \alpha r, \quad \alpha = \begin{pmatrix} -1 & -1 & 0 & 0 \\ 1 & 0 & 0 & -1 \\ 1 & -1 & 1 & -1 \\ 0 & 1 & -1 & 0 \end{pmatrix}, \quad r = \begin{pmatrix} k_1 y_1 \\ k_2 y_1 y_3 \\ k_3 y_4 \\ k_4 y_2 y_3 \end{pmatrix}. \tag{10.2}$$

The columns of the matrix α are the stoichiometric vectors for the individual reactions and the components of r are the corresponding rates. The k_i are the reaction rate constants. We study the system with

$$k^T = (10^{-9}, 10^7, 10^3, 10^9) \text{ and } y^T(0) = (2 \times 10^{-3}, 0, 0, 0).$$

The Jacobian of the system is

$$J = \begin{pmatrix} -(k_1 + k_2 y_3) & 0 & -k_2 y_1 & 0 \\ k_1 & -k_4 y_3 & -k_4 y_2 & 0 \\ k_1 - k_2 y_3 & -k_4 y_3 & -(k_2 y_1 + k_4 y_2) & k_3 \\ k_2 y_3 & 0 & k_2 y_1 & -k_3 \end{pmatrix},$$

which is singular, corresponding to the invariant $y_2 - y_3 - y_4$. This does not affect the computations to any great extent although we could have chosen to reduce the order of the system at the expense of complicating the right-hand sides. For the standard integration formulae which are linear in function and derivative values Robertson and McCann (1969) have pointed out that the approximate solutions also satisfy the invariant relation apart from round-off. We note that y_2, y_3, y_4 are intermediates, and that y_3 and y_4 are very reactive species as shown by the magnitudes of the rate constants (k_2, k_3, k_4) of reactions 2, 3 and 4 in which they participate as reactants. We identify the fast motions with y_3 and to a lesser extent with y_4. Typical values of the variables are $y^T = (2 \times 10^{-3}, .4 \times 10^{-10}, .2 \times 10^{-11}, .3 \times 10^{-10})$. y_1 is a very unreactive species as indicated by the magnitude of k_1. y_2 is a factor of 10^{-2} more reactive but still very much less active than y_3 or y_4.

In this system we are able to provide a physical justification of the structure of the Jacobian in terms of reactive species and the corresponding fast reactions in which they participate as reactants and insertion of numerical values in the Jacobian confirms this relationship.

Turning now to the Jacobian of the nonlinear equations F = 0 in its partitioned form $\begin{pmatrix} a & b \\ c & d \end{pmatrix}$, we observe that a is almost diagonal, has much smaller elements and is much more slowly varying than b,c, or d. We therefore assume that a can be updated less frequently than b,c,d. Four separate runs of the programme at tabular interval 10^3 and over the range $0 - 10^5$ have been carried out with different updating strategies. A corresponds to use of a = I throughout and therefore gives a Newton-like iteration for the fast motions and repeated substitution for the slow motions as in (3.6). In B, a = I initially and thereafter a is updated and a full inversion of the Jacobian is carried out whenever the perturbation E_{11} corresponding to a satisfies $||E_{11}||_\infty > \frac{1}{2}$. C is similar to B but the criterion for updating is $||E_{11}||_\infty > \frac{1}{2}\, 10^{-1}$. In D the full Jacobian is updated and inverted when convergence is unsatisfactory. Thus whenever we refer to an inversion in A,B, & C it means an update of the inverse with respect to changes in the fast motions only, the recomputation of a full inverse only being called for whenever a is updated.

We have taken iteration (NIT), inverse (NINV) and a-update counts (UPD) for the tabular intervals $x = 0\ (10^3)10^4$ and the results are shown in Table 1. The programme uses an explicit integration method to begin with and then switches to the trapezoidal rule. Automatic step adjustment takes place in order to satisfy a local relative accuracy criterion of 10^{-3}. In all cases a large number of iterations and inversions take place in the first interval where the step length is initially small and builds up by several magnitudes to $h = 10^2$. Clearly, in this region a = I is almost as good an approximation as using the full Jacobian. Over the first three intervals, methods A and B are very similar and methods C and D are very similar in performance. Thereafter the step-length in A becomes limited by our convergence restriction which allows a maximum of four iterations per step. The inversions here are associated with attempts to increase the interval, an unsuccessful doubling of the interval being followed immediately by halving. In B, C and D convergence remains satisfactory and only one inversion takes place for C and D associated with interval increase. The approximate inverses are all different constants over this range.

TABLE 1

INVERSION AND ITERATION COUNTS

	A		B			C			D	
$10^{-3}x$	NIT	NINV	NIT	NINV	UPD	NIT	NINV	UPD	NIT	NINV
1	816	27	816	27	0	812	27	1	796	27
2	26	2	33	3	1	23	2	0	22	2
3	19	1	10	0	0	18	1	0	18	1
4	23	2	8	0	0	10	1	0	8	1
5	23	2	8	0	0	8	0	0	8	0
6	23	2	8	0	0	8	0	0	8	0
7	23	2	10	0	0	8	0	0	8	0
8	23	2	8	0	0	10	0	0	8	0
9	23	2	8	0	0	8	0	0	8	0
10	23	2	8	0	0	8	0	0	8	0
TOTAL	1022	44	917	30	1	913	31	1	892	31

The table shows the numbers of iterations, inversions and full matrix updates performed in the tabular intervals as shown. In cases A, B and C NINV refers to the number of partial inversions whereas in case D it refers to inversion of the full matrix. UPD gives the corresponding number of updates of the partition of the Jacobian associated with the slow motions (the matrix a of (3.1)). The updating strategies are

A. $a = I$

B. $||E_{11}|| < \frac{1}{2}$

C. $||E_{11}|| < \frac{1}{2}\ 10^{-1}$

D. full matrix update

E_{11} is the perturbation associated with the slow motions and a maximum of four iterations is allowed for convergence at each step. Further computation over the range $10^4 \leqslant x \leqslant 10^5$ shows that the step-length of A continues to be restricted by slow convergence. About ten more inversions are required for C and D, each of them full inversions. A similar number of inversions is required for B but only three of them are full inversions.

To summarise, we observe that the convergence properties of A are satisfactory up to $x = 3 \times 10^3$ and thereafter B gives practically the same iteration count as D. If A is used over the first part of the range and then we switch to B, only one full inversion and 30 "partial" inversions are required compared with 31 full inversions for D. In actual fact, due to the special structure of a, this partition may be adequately represented by a diagonal matrix throughout so that a full inverse need never be computed.

This example illustrates the potential improvements in computational efficiency to be gained by taking account of the structure of the perturbation E in matrix updating strategies. The improvements will depend in general on the range of integration, the stiffness and structure of the differential equations, convergence and step adjustment criteria. This system of equations is very stiff with a good separation between the eigenvalues associated with slow and fast motions. Under these circumstances the approximation $a = I$ is well worthwhile. This leads to the iteration (3.6) corresponding to a Newton-like iteration for the fast motions and a repeated substitution with corrections for the slow motion. Periodic updating of a (reduction of E_{11} to zero) leads to further efficiency.

CONCLUSION

The efficiency of numerical methods for the solution of stiff differential equations with special structure is enhanced by the use of appropriate iterative methods for the solution of the implicit integration equations and suitable methods of inverse updating.

14

PARABOLIC PARTIAL DIFFERENTIAL EQUATIONS

1. INTRODUCTION

In this chapter various methods of solving the differential equation

$$\frac{\partial u}{\partial t} = \sum_{i,j=1}^{m} \frac{\partial}{\partial x_j}\left(k_{ij} \frac{\partial u}{\partial x_j} \right) + f(u,\underset{\sim}{x},t), \quad (m=1,2 \text{ or } 3) \tag{1.1}$$

in some region, $R\times[a,b]$, subject to the initial condition

$$u(\underset{\sim}{x},a) = u_0(\underset{\sim}{x}), \quad (x\in R)$$

together with boundary conditions for $\underset{\sim}{x} \in \partial R$ are given. In this definition it is assumed that $\underset{\sim}{x}=(x_1,\ldots,x_m)^T\in R$ an open and bounded set having a closed boundary denoted by ∂R.

In the absence of convection terms (1.1) is the general form for any equation representing the diffusion of heat or mass. If $k_{ij}=k_{ij}(u)$, as for example when u is temperature, and the thermal conductivity varies with the temperature, the equation is nonlinear. If the diffusivity varies through R then $k_{ij}=k_{ij}(\underset{\sim}{x})$, whereas if it is constant the equation may reduce to the simple form

$$\frac{\partial u}{\partial t} = k \sum_{i=1}^{m} \frac{\partial^2 u}{\partial x_i^2} + f(u).$$

In the following section various methods will be outlined that reduce (1.1) to a system of ordinary differential equations of the form:

$$B\underset{\sim}{y}' = A(\underset{\sim}{y}) + \underset{\sim}{f}(\underset{\sim}{y}), \tag{1.2}$$

where the matrix B is not necessarily the identity matrix.

2. PIECEWISE POLYNOMIALS

Probably the most popular method of carrying out the reduction of (1.1) to the system (1.2) is by one of the semi-discrete forms of the finite element method (Mitchell and Wait, 1976) in which a separable approximation of the form

$$U(\underset{\sim}{x},t) = \sum_i y_i(t)\, \varphi_i(\underset{\sim}{x}) \tag{2.1}$$

is obtained. The basis functions φ_i are taken as piecewise polynomials defined on the region R partitioned into small elements by means of a mesh (see figure 1). Such functions have low global continuity in the sense that they might be continuous functions but have discontinuities in first derivatives on the element boundaries, or possibly they could be C^1 functions with discontinuities in the second derivatives. Such piecewise polynomial functions will be mentioned later in Chapter 19 during the discussion of Galerkin methods for boundary value problems. It should be emphasised that if for example the basis function $\varphi_i(x)$ is the piecewise linear function defined by

$$\varphi_i(x) = \begin{cases} \dfrac{x-\xi_{i-1}}{\xi_i-\xi_{i-1}}, & x \in [\xi_{i-1},\xi_i] \\[2ex] \dfrac{x-\xi_{i+1}}{\xi_i-\xi_{i+1}}, & x \in [\xi_i,\xi_{i+1}] \\[2ex] 0, & \text{otherwise}, \end{cases}$$

then it follows that

$$y_i(t) = U(\xi_i,t), \qquad (t \geq a;\ i=1,\ldots,N). \tag{2.2}$$

Similarly the piecewise cubic functions shown in figure 1 lead to 2N unknown functions $y_i(t)$ $(i=1,\ldots,2N)$ such that

$$y_{2i-1}(t)=U(\xi_i,t)$$

and $\qquad (t\geq a; j=1,\ldots,N) \qquad (2.3)$

$$y_{2i}(t) = \frac{\partial U(\xi_i,t)}{\partial x},$$

assuming the basis functions are ordered appropriately. There is not room in this short section to enumerate the myriad forms of piecewise polynomials; the interested reader is referred to Mitchell and Wait (1976) where an extensive catalogue of the possible alternatives will be found. Only the two most popular methods, namely Galerkin and Collocation will be considered here, but other variations such as Least Squares are possible (ibid.).

3. GALERKIN

This form of the Galerkin method, alternatively called Kantorovich-Galerkin or Faedo-Galerkin (Strang and Fix 1973) seeks to find an approximate solution of the weak form of equation (1.1), that is

$$\left(\frac{\partial U}{\partial t}, \varphi_l\right) = -\sum_{i,j=1}^{m} \left(k_{ij}\frac{\partial U}{\partial x_j}, \frac{\partial \varphi_l}{\partial x_i}\right) + (f,\varphi_l), \qquad (3.1)$$

for each basis function φ_l, where

$$(u,v) = \iint_R u(\underset{\sim}{x})v(\underset{\sim}{x})\, d\underset{\sim}{x}$$

and it is assumed that the boundary condition

$$u(\underset{\sim}{x},t)=0, \qquad (t\in[a,b])$$

is satisfied by all the basis functions. The initial condition corresponding (3.1) is

$$(U(0),\varphi_l) = (u_0,\varphi_l), \quad (\text{for all } \varphi_l). \qquad (3.2)$$

Thus if the approximate solution is given by (2.1) it follows from (3.1) and (3.2) that the functions $y_i(t)$ satisfy (1.2) where $B = \{b_{pq}\}$ with

$$b_{pq} = (\varphi_p, \varphi_q); \tag{3.3}$$

$$A(\underset{\sim}{y}) = D(\underset{\sim}{y}).\underset{\sim}{y} \text{ where } D=\{d_{pq}\}$$

such that

$$d_{pq} = -\sum_{ij=1}^{m} \left(k_{ij} \frac{\partial\varphi_q}{\partial x_j}, \frac{\partial\varphi_p}{\partial x_i}\right) \tag{3.4}$$

and

$$\underset{\sim}{f}(\underset{\sim}{y}) = \{(f,\varphi_l)\}. \tag{3.5}$$

The initial conditions are then obtained from (3.2) as

$$B\underset{\sim}{y}(0)=\underset{\sim}{g}$$

where $g=\{(u_0,\varphi_l)\}$. Observe than an approximation procedure defined by (3.1) has two important features:

a) The equations involve integrations over the region R. Although this is reduced to integrations over a small number of elements, as φ_l is identically zero in all but a few elements, any step-by-step method would require the right hand side of (1.2) to be re-evaluated at each step. If f is nonlinear, this involves integrals of the form

$$\iint_R f\left(\sum_j y_i(t_n)\varphi_j(\underset{\sim}{x})\right) \varphi_l(\underset{\sim}{x})d\underset{\sim}{x}.$$

This is time consuming and it has been found in practice (Hopkins 1975) that when a variable-step stiff-equation solver is used to provide the solution, most of the time is spent in computing these integrals.

b) The approximate solution to a second order equation (1.1), when defined by (3.1) need only have first derivatives that are square integrable, that is such that

b) cont.

$$\iint_R \frac{\partial\varphi_p}{\partial x_i} \frac{\partial\varphi_q}{\partial x_j} d\underset{\sim}{x} = \sum_{s=1}^{S} \iint_{R_s} \frac{\partial\varphi_p}{\partial x_i} \frac{\partial\varphi_q}{\partial x_j} d\underset{\sim}{x}$$

exists and is finite, where it is assumed that the region R is partition exactly into non-overlapping elements $R_s(s=1,...,S)$.

4. CONVERGENCE OF GALERKIN APPROXIMATIONS

There follows in this section a brief outline of the a priori convergence estimates that are obtained for continuous Galerkin approximations. That is, for any $t\in[a,b]$, there exists a bound of the form

$$||u-U||_{L_2(R)} \leqslant C_1 h^{\alpha}, \tag{4.1}$$

where

$$||u||^2_{L_2(R)} = (v,v)$$

and h is a measure of the refinement of the mesh, say in one dimension

$$h = \max_j (\xi_j - \xi_{j-1}).$$

The integer α is a measure of the accuracy of approximation by the functions φ_j; namely that it is possible to interpolate exactly polynomials of degree $\alpha-1$ with a function of the form

$$W(\underset{\sim}{x}) = \sum_j \beta_j \varphi_j(\underset{\sim}{x}).$$

The proof is straightforward, provided $k_{ij} \neq k_{ij}(u)$ and that f is Lipschitz continuous, that is

$$|f(u) - f(v)| \leqslant C_2 |u-v|.$$

The interested reader is referred to Thomée and Wahlbin (1975).

5. COLLOCATION

It was pointed out in Section 3 that two of the most significant features of Galerkin approximations are the need to evaluate integrals and the low order of continuity required by the basis functions. Collocation on the other hand requires no evaluation of integrals, but does require the basis functions to have (piecewise) continuous second derivatives, in the case of second order equations. Thus it is not possible to use the piecewise linear functions in collocation, the functions invariably used are piecewise cubic (see figure 1).

The method of collocation is to obtain an approximate solution of the form (2.1) that satisfies the differential equation (1.1) at a finite number of collocation points. If the approximation is defined in terms of 2N piecewise cubic basis functions then 2N collocation points are required. These points are chosen to be ξ_1 and ξ_N (to satisfy the boundary conditions) together with two points ζ_{2i}, ζ_{2i+1} in each subinterval (ξ_i, ξ_{i+1}). The position of these points is critical if the convergence estimates are to have the same exponent of h as (4.1). A poor choice of collocation points can also lead to numerical instabilities in the computation and so it is of more than academic interest to find the appropriate positions. Douglas and Dupont (1973) have shown that the correct positions for the internal collocation points are

$$\zeta_{2i+k} = \frac{1}{2}(\xi_i+\xi_{i+1}) - \frac{(-1)^k}{2\sqrt{3}}(\xi_{i+1}-\xi_i), \qquad (k=0,1)$$

that is the points are the zeroes of the quadratic Legendre polynomial on the interval $[\xi_i, \xi_{i+1}]$. Douglas and Dupont have further shown (1974) that in general if piecewise polynomials of degree r are used then the zeroes of Legendre polynomials of degree r-1 should be taken as the collocation points.

The approximation with a piecewise cubic basis is then defined for $t \in [a,b]$

$$\left[\frac{\partial U}{\partial t} - \frac{\partial}{\partial x}\left(k\,\frac{\partial U}{\partial x}\right) - f(U,x,t)\right] = 0, \quad \begin{cases} i=1,\ldots,N-1 \\ x=\zeta_{2i+k};\ k=0,1 \end{cases} \tag{5.1}$$

together with the boundary conditions satisfied at $x=\xi_1$ and ξ_N.

Thus it follows that the functions $y_j(t)$ satisfy (1.2) where (c.f. (3.3)-(3.5))

$$b_{pq} = \varphi_q(\zeta_p), \tag{5.2}$$

$$d_{pq} = \left[\left(\frac{\partial}{\partial x}\,k\,\frac{\partial\varphi_q}{\partial x}\right)\right]_{x=\zeta_p} \qquad \begin{pmatrix} p=2,\ldots,2N-1 \\ q=1,\ldots,2N \end{pmatrix} \tag{5.3}$$

and

$$\underset{\sim}{f} = \{f(U,\zeta_p,t)\}, \tag{5.4}$$

with the first and last rows determined by the form of the boundary conditions.

Douglas and Dupont (1974) have derived a priori convergence estimates of the form

$$||u-U||_{L_\infty(R)} \leqslant C_3\,h^\alpha \tag{5.5}$$

for this type of collocation, where

$$||v||_{L_\infty(R)} = \sup_{\underset{\sim}{x}\in R\cup\partial R} |v(\underset{\sim}{x})|.$$

It can be shown that both the Galerkin and Collocation convergence estimates, (4.1) and (5.5) respectively, can be improved at the knots of the mesh, this is the so called <u>superconvergence</u> of these methods. Recent numerical tests (Hopkins 1975) indicate that there is not a great difference here in the numerical accuracy of the two methods when combined with a stiff-equation solver. It should be noted that in both cases the matrices B and D are sparse and banded, and this can be used to achieve a highly efficient numerical procedure.

6. THE METHOD OF LINES

This is a method that has been studied extensively in the Russian literature (for example, Berezin and Zhidkov 1965) but has been almost totally ignored elsewhere. The partial differential equation (1.1) is reduced to a system of ordinary differential equations by simply replacing the spatial derivatives ($\frac{\partial}{\partial x_i}$) by differences. As a particular example, the one dimensional diffusion equation considered in section 5 could be replaced by

$$\frac{dy_i}{dt} = \frac{1}{h^2}\{k(\xi_{i+1/2})(y_{i+1}-y_i)-k(\xi_{i-1/2})(y_i-y_{i-1})\}+f(y_i,\xi_i,t), \quad (6.1)$$

$$(i=1,\ldots,N-1)$$

together with the boundary conditions satisfied at $x=\xi_0$ and ξ_N. The validity of such difference replacements will be discussed at greater length in the chapters on Boundary value problems, but clearly it is assumed that $y_i(t)$ is an approximation to $u(\xi_i,t)$. Again the functions $y_i(t)$ satisfy a system of equations given by (1.2) but now B=I, while D and $\underset{\sim}{f}$ also have a very simple form.

In the next chapter Numerov's method will be discussed. In this method the equation

$$y''(x) = f(x),$$

is replaced by the difference equation

$$\frac{1}{h^2}[y_{i-1}-2y_i+y_{i+1}] = \frac{1}{12}[f(x_{i-1})+10f(x_i)+f(x_{i+1})].$$

A similar procedure has been suggested (Zafarullah 1970) to provide a more accurate formulation of the method of lines. That is (6.1) is replaced by (assuming k=constant)

$$\frac{1}{12}\left\{\frac{dy_{i-1}}{dt}+10\frac{dy_i}{dt}+\frac{dy_{i+1}}{dt}\right\} = \frac{k}{h^2}\cdot\{y_{i-1}-2y_i+y_{i+1}\}$$

$$+\frac{1}{12}\{f(\xi_{i-1},y_{i-1})+10f(\xi_i,y_i)+f(\xi_{i+1}y_{i+1})\}. \quad (6.2)$$

If $\underset{\sim}{u}(t) = \{u(\xi_i,t),\ i=1,\ldots,N\}$ then Zafarullah has shown that

$$||\underset{\sim}{u}(t) - \underset{\sim}{v}(t)|| \leq C_4\ h^{7/2},$$

where

$$||\underset{\sim}{v}|| = \{\sum_{i=1}^{N} v_i^2\}^{1/2}.$$

Thus the method should be competitive with the use of piecewise cubic polynomials in Collocation or Galerkin methods, however the results of numerical tests are not encouraging (Wait and Hopkins 1975).

7. CONCLUSION

The main advantage of converting a single partial differential equation into a system of stiff ordinary differential equations is that the sophisticated algorithms available lead to an efficient variable step method of solving the p.d.e. which keeps some control over error growth. The same cannot be said for standard finite difference replacements such as the Crank-Nicolson method. Attempts have been made to implement variable step finite difference methods but the step control mechanism tends to be crude and rather inefficient.

An example of a practical implementation of the method of lines together with computer procedures can be found in Sincovec and Madsen (1975).

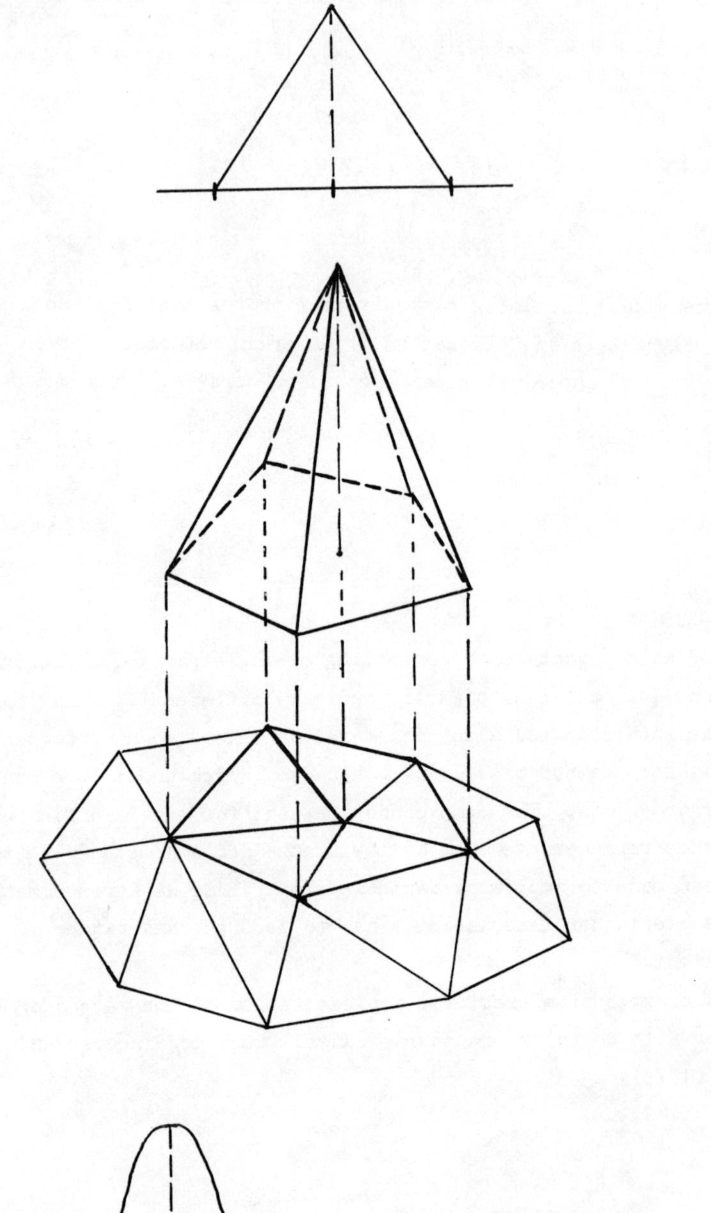

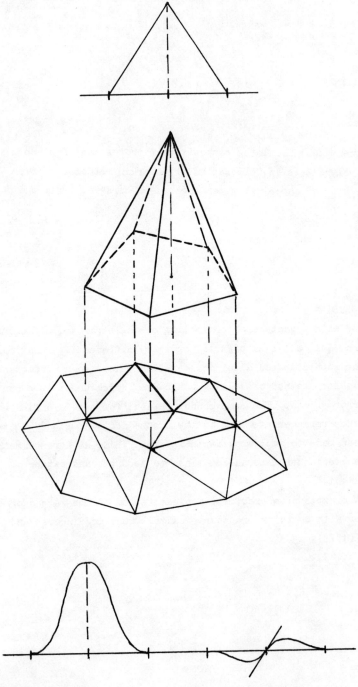

Figure 1

PART 3

BOUNDARY VALUE PROBLEMS

15

BOUNDARY VALUE PROBLEMS; FINITE DIFFERENCE METHODS

1. INTRODUCTION

Boundary value problems frequently arise in engineering and science, often as the result of separating variables in a partial differential equation. Whilst the difference between initial value problems and problems where the boundary conditions are given at two or more points may appear slight, there is a fundamental difference in their difficulty. In Roberts and Shipman (1972) some examples are given (p.2) which demonstrate this point.

For initial value problems the theory is reasonably well developed but for boundary value problems the theory is deficient in may respects. The deficiency usually manifests itself in the proofs giving sufficient conditions but not necessary conditions.

In general we classify the method of solution for boundary value problems into three basic categories

i) shooting methods,

ii) finite difference methods,

iii) variational, collocation and others.

Keller (1975b) has written a valuable survey paper covering all these techniques. In this chapter we will restrict ourselves to category ii); succeeding chapters will cover the other methods. This chapter will also omit a discussion of the problems that arise in the presence of singularities.

We note that in general we may have a differential equation of order m, in which case we must have m boundary conditions distributed in some way between at least two points and in general we will assume that they apply at the two end points.

Further we may reduce any high order differential equation to a system of first order equations by the usual method (see Chapter 1 and Roberts & Shipman (1972) p.8).

2. BASIC METHODS

The application of finite difference methods to boundary value problems is basically straightforward. Any derivatives occurring in the differential equation are replaced by their finite difference approximations.

As a first example consider the boundary value problem

$$\begin{aligned} &y'' - f(x,y,y') = 0, \\ &y(a) = \alpha, \quad y(b) = \beta \end{aligned} \tag{2.1}$$

If we take a uniform mesh, that is take

$$x_n = a + nh, \quad 0 \leq n \leq N, \quad h = (b-a)/N, \tag{2.2}$$

then a simple difference approximation gives rise to the equations

$$\begin{aligned} &(y_{n+1} - 2y_n + y_{n-1})/h^2 - f(x_n, y_n, (y_{n+1} - y_{n-1})/2h) = 0 \\ &y_0 = \alpha, \quad y_N = \beta, \quad 0 < n < N . \end{aligned} \tag{2.3}$$

The approximate solution is thus obtained by solving this system of (N+1) equations in (N+1) unknowns.

If f in (2.1) does not depend on y', a more accurate approximation is given by Numerov's method which uses

$$\begin{aligned} &(y_{n+1} - 2y_n + y_{n-1})/h^2 - [f(x_{n+1},y_{n+1}) + 10f(x_n,y_n) + f(x_{n-1},y_{n-1})]/12 \\ &\qquad = 0 , \quad (2.4) \\ &y_0 = \alpha, \quad y_N = \beta, \quad 0 < n < N. \end{aligned}$$

As a final example, consider a system of s first order equations

$$y' - f(x,y) = 0$$

with boundary conditions (2.5)

$$g(y(a), y(b)) = 0.$$

Here the values of y, f and g are s-component vectors. An obvious simple approximation that has been studied in detail by Keller (1969, 1974) is

$$\begin{aligned} &(N_h y)_n := (y_n - y_{n-1})/h - f(x_{n-\frac{1}{2}}, (y_n + y_{n-1})/2) = 0, \\ &g(y(a), y(b)) = 0, \qquad 0 < n \leq N. \end{aligned} \tag{2.6}$$

The discretization error of this approximation is defined to have components

$$d_0 = g(y(a), y(b)),$$

$$d_n = \left\{y(x_n) - y(x_{n-1})\right\}/h - f(x_{n-\frac{1}{2}}, \left\{y(x_n) + y(x_{n-1})\right\}/2) \qquad (2.7)$$

$$0 < n \leq N.$$

It is stable if constants K and h_0 exist such that for all $h \leq h_0$ and for all $y = \{y_0, \ldots, y_N\}$ and $z = \{z_0, \ldots, z_n\}$ in some suitable region, and with

$$\|y - z\| = \max_{0 \leq n \leq N} \|y_n - z_n\|,$$

we have

$$\|y - z\| \leq K \max_{0 < n \leq N} \|(N_n y)_n - (N_h z)_n\|.$$

The definitions of these concepts for other discretizations are analogous to these and are given by Keller and White (1975), who give a general theory of difference methods for boundary value problems (see also Keller (1975a)). The main result of this work is:

For isolated solutions of a non-linear problem the difference scheme has, for sufficiently small h, a unique solution converging to the exact solution if

i) the linearised difference equations are stable and consistent for the linearised problem,

ii) the linearised difference operator is Lipschitz continuous,

iii) the non-linear difference equations are consistent with the non-linear differential equation.

Newton's method is shown to be valid, with quadratic convergence, for computing the numerical solution from a sufficiently close approximation, for sufficiently small h.

They also show, as did Kreiss (1972), that the difference approximation is stable if it is "as compact as possible". An approximation is "as compact as possible" if each equation contains only (m+1) of the y_i when the highest derivative occurring in the differential equation is of order m. Thus (2.3), (2.4) and (2.7) are as compact as possible, and hence stable; they are consistent of orders two, four and two respectively and hence convergent of these orders. On the other hand the obvious fourth order consistent approximation to $y'' - f(x,y) = 0$ that is

$$(-y_{n+2} + 16y_{n+1} - 30y_n + 16y_{n-1} - y_{n-2})/12h^2 - f(x_n, y_n) = 0,$$

is not as compact as possible.

We illustrate some of these concepts by discussing (2.1) with discretization (2.3). The linearised form of (2.1) is

$$\begin{aligned} &z'' - p(x)\, z' - q(x) z = r(x), \\ &z(a) = \alpha_1, \qquad z(b) = \beta_1, \end{aligned} \tag{2.8}$$

where

$$\begin{aligned} p(x) &= \frac{\partial f}{\partial y'}\,(x,\, y(x),\, y'(x)), \\ q(x) &= \frac{\partial f}{\partial y}\,(x,\, y(x),\, y'(x)). \end{aligned} \tag{2.9}$$

The linearised difference equations are

$$\begin{aligned} &a_n z_{n-1} + b_n z_n + c_n z_{n+1} = -h^2 r_n, \\ &z_0 = \alpha_1, \qquad z_n = \beta_1, \end{aligned} \tag{2.10}$$

where

$$\begin{aligned} &a_n = -(1 + \tfrac{h}{2}\, p(x_n)), \quad b_n = 2 + h^2 q(x_n), \quad c_n = -(1 + \tfrac{h}{2}\, p(x_n)), \\ &r_n = r(x_n) \qquad 0 < n < N. \end{aligned} \tag{2.11}$$

If now $P^* = \max_{a \leq x \leq b} |p(x)|, \qquad q(x) \geq Q_* > 0, \quad q \leq x \leq b$

and $\qquad h \leq 2/P^*$,

it can be shown in an elementary way that (2.10) is stable (Keller (1968) Isaacson and Keller (1966)). In this case also the triple-diagonal equations (2.10) are diagonally dominant, and are easily and efficiently solved accurately.

3. SOLUTION OF THE ALGEBRAIC EQUATIONS

If the differential equation and boundary conditions are linear, so are the equations of the discretization. For example (2.8) gives the triple-diagonal set of equations (2.10), and the matrix of these equations displays the typical structure obtained from finite difference methods. If a low order approximation is used the matrix will have a narrow band-width, but for good accuracy h must be small and the matrix will have large dimensions. For a high order approximation it will often be possible to use a smaller matrix, but its bandwidth will be higher and one will have to investigate the stability carefully.

The simple structure of the resultant equations is one of the principle attractions of the finite difference method, so that if unequal mesh-widths are used, or special approximations are used at the boundaries, care should be taken to retain this simplicity. Keller (1974) discusses in some detail efficient ways of solving the linear equations by block and band methods.

For non-linear equations a modified Newton's method is usually used to solve the discrete equations. We can write this in the form

$$J_t(\underline{y}^t - \underline{y}^{t-1}) = -F(\underline{y}^{t-1}), \tag{3.1}$$

where the function F has N components coming from the approximations to the system of differential equations (2.5) and one component from the boundary conditions, and in the case of a system of s differential equations, each of these (N+1) components itself has s components. In practice the Jacobian J_t of F, which should be evaluated at $\underline{y}^{t-1}$, will not be re-evaluated at each iteration, but only when necessary to ensure convergence. That is, if the iteration is converging with a fixed J, the time-consuming process of re-evaluated and triangularising J is omitted.

The matrix J is square of order (N+1)s. It is of block band form with square blocks of order s; these blocks will in general involve the Jacobian matrix $J_f(x_n, y_n^{t-1})$ of the function f occurring in the differential equations.

As a simple example we again revert to (2.3), for which each iteration of the solution by Newton's method involves the solution of the set of linear equations

$$\begin{aligned} a_n e_{n-1}^t + b_n e_n^t + c_n e_{n+1}^t &= -(N_h \underline{y}^{t-1})_n, \\ e_0 = e_N = 0, \qquad & 0 < n < N, \end{aligned} \tag{3.2}$$

where

$$e_n^t = y_n^t - y_n^{t-1},$$

and a_n, b_n and c_n are given by (2.11) and (2.9) with the derivatives of f evaluated at

$$(x_n, y_n^{t-1}, (y_{n+1}^{t-1} - y_{n-1}^{t-1})/2h).$$

As always with Newton's method the speed of convergence depends on a good approximation to the solution being available to start the iteration.

4. EXTRAPOLATION

Two well known techniques are available for increasing the order of a finite difference method and thus improving the accuracy.

We first describe extrapolation with reference to the discretization (2.3) of problem (2.1). The basis of the method is the result

$$y_n = y(x_n) + h^2 e(x_n) + O(h^4), \quad 0 \leq n \leq N, \tag{4.1}$$

where

$$\begin{aligned} e''(x) - p(x)e'(x) - q(x)e(x) - r(x) &= -\frac{h^2}{12}(-y^{1V}(x) + p(x)y'''(x)), \\ e(a) = e(b) &= 0 \end{aligned} \tag{4.2}$$

and p, q, and r are given by (2.9).

On evaluating the approximate solution at steplength h = k and h = k/2 using (2.3), we find by means of (4.1), in an obvious notation,

$$\tfrac{4}{3}(y_{2n,k/2} - y(x_n)) - \tfrac{1}{3}(y_{n,k} - y(x_n)) = O(h^4) \tag{4.3}$$

and an improved solution with error $O(h^4)$ is

$$\bar{y}_n = \frac{4}{3}\, y_{2n,k/2} - \frac{1}{3}\, y_{n,k} \qquad 0 \leq n \leq N.$$

This technique has been discussed in detail in Chapter 7 in connection with the initial value problem. The extrapolation can be repeated, using solutions at more than two step-lengths, provided that the coefficients of the differential equations are sufficiently smooth.

Keller (1974) has shown that extrapolation can be used with the modified Euler method (2.7) for first order systems even if f(x,y) or its derivatives have discontinuities at given values of x, or if the mesh spacing is unequal, provided that the steplength at any point x is given by $\theta(x)h$ where $\theta(x)$ is piecewise constant and that the points of discontinuity of f(x,y) and of $\theta(x)$ are mesh-points of every mesh.

When using this method, a good starting approximation for the solution of the non-linear algebraic equations for the solution on a fine mesh is available by interpolation from the previously calculated solution on a coarser mesh.

5. DEFERRED CORRECTION

An alternative approach is provided by Fox's method of deferred correction, which has recently been put on a sound theoretical basis by Pereyra (1966, 1967, 1968). A Fortran code for the boundary value problem for first order systems has recently been published by Lentini and Pereyra (1974, 1975). For an elementary treatment see Watt (1968).

The theoretical basis for the method also depends on the existence of an asymptotic error expansion (see (4.1) for the first term), but with extra iterations the order of the method is preserved if an expansion does not exist. The basis of deferred correction is that an accurate finite difference operator is split into two parts, the first a simple stable part such as (2.3) or (2.6) and the second a small part, of order h^2, h^4 and h^2 respectively for those methods. The solution is then obtained by the obvious iteration.

We illustrate the ideas by means of (2.1), but we omit the dependence of f on y' merely to simplify the description. Using the approximation

$$y''(x_n) = \frac{1}{h^2}\left\{\delta^2 - \frac{1}{12}\delta^4 + \frac{1}{90}\delta^6 - \frac{1}{560}\delta^8 + \dots\right\}y(x_n), \tag{5.1}$$

where δ is the central difference operator, we have

$$\frac{1}{h^2}\left\{\delta^2 - \frac{1}{12}\delta^4 + \frac{1}{90}\delta^6 - \frac{1}{560}\delta^8 + \dots\right\}y(x_n) - f(x_n,y(x_n)) = 0 ,$$

or

$$\left\{y(x_{n+1}) - 2y(x_n) + y(x_{n-1})\right\}/h^2 - f(x_n,y(x_n))$$

$$= \frac{1}{h^2}\left\{\frac{1}{12}\delta^4 - \frac{1}{90}\delta^6 + \frac{1}{560}\delta^8 - \dots\right\}y(x_n).$$

The usual approximation (c.f.(2.3),) is obtained by neglecting the right hand side and solving the resulting equations for $\{y_n\}$. The method of deferred correction however calculates in succession the approximate solutions $\{y_n^0\}$, $\{y_n^1\}$, $\{y_n^2\}$, ... using the equations

$$\begin{aligned} &(y_{n+1}^r - 2y_n^r + y_{n-1}^r)/h^2 - f(x_n,y_n^r) = (Cy^{r-1})_n, \qquad 0 < n < N , \\ &y_0^r = \alpha, \qquad\qquad y_N^r = \beta. \end{aligned} \tag{5.2}$$

Here $(Cy^0)_n = 0$ and if n is not too close to 0 or N

$$(Cy^{r-1})_n = \frac{1}{h^2}\left\{\frac{1}{12}\,\delta^2 - \frac{1}{90}\,\delta^4 + \frac{1}{560}\,\delta^8 - \dots\right\} y_n^r . \qquad (5.3)$$

If n is close to 0 or N, the calculation of $(Cy^{r-1})_n$, that is of the nth component of the result of applying the operator C to y^{r-1}, requires values of y_q^{r-1} which are outside the range $0 \leq q \leq N$. For this case equivalent (unsymmetric) formulae using only the available values of y_q^{r-1} must be used in place of (5.3). In his work Pereyra has used automatic methods to derive the coefficients in such approximations.

Notice that for this approximation, C is of order h^2 and it can be shown that $\{y_n^r\}$ has error of order h^{2r+2}, and these results still hold if in calculating C only differences up to δ^{2r+2} are used.

If the equations are linear, their solution $\{y_n^r\}$ for $r > 0$ is easy, since a triangular factorisation of the matrix is already available. In the non-linear case, when the modified Newton method is being used, a sufficiently accurate Jacobian and its factorisation may also be available from the solution of the equations for a previous value of r.

16

SHOOTING METHODS FOR BOUNDARY VALUE PROBLEMS

1. DESCRIPTION OF THE BASIC METHOD

Suppose we have a system of ordinary differential equations with boundary conditions given at two or more points of the range. Assuming the equations are in the standard first-order form, we can write them as

$$\left.\begin{aligned} &\frac{dy_i}{dx} = f_i(x, y_1, y_2, \ldots, y_s), \quad i = 1,2,\ldots,s, \\ \text{or} \\ &\underline{y}' = \underline{f}(x, \underline{y}) \,. \end{aligned}\right\} \tag{1.1}$$

For s first-order equations we need s boundary conditions to define a solution. The general form of boundary condition is a nonlinear relation between the values of $y_i(x)$ at a number of points $x = a, b, c, \ldots$. However, for most of our discussion we will take a much simpler form, and assume that s boundary values of the components $y_i(x)$ are specified, some at a and some at b, with the solution required in the range (a,b). Let the boundary conditions be

$$\left.\begin{aligned} y_i(a) &= \alpha_i, \quad i = 1,2,\ldots k, \\ y_i(b) &= \beta_i, \quad i = i_{k+1},\ldots,i_s, \end{aligned}\right\} \tag{1.2}$$

so that we have k boundary conditions at x = a, and s − k conditions at x = b, where $0 < k < s$. (The indices $i_{k+1}, i_{k+2}, \ldots, i_s$ are distinct but need not be different from the integers $1,2,\ldots k$.)

A great many practical examples of boundary-value problems may be written in the form of (1.1) and (1.2). Two important extensions which will be considered later are the case of an infinite range, $b \to \infty$ say, and the case of periodic boundary conditions

$$y_i(a) = y_i(b), \quad i = 1,2,\ldots,s. \tag{1.3}$$

Returning to (1.1) and (1.2), the differential equations may be solved by initial-value techniques if we can find enough boundary values at $x = a$ or $x = b$, i.e. if we can obtain values for all the s variables $y_i(x)$ at a or b. Suppose we estimate the values of $y_i(x)$ which are not already specified by (1.2); then we have enough information to integrate the system forwards from $x = a$ and backwards from $x = b$. If our estimates are correct, the forward and backward solutions will be equal at intermediate points of the range. The problem therefore is to improve the initial estimates of the unknown boundary conditions until the two solutions match, and this is the basis of the shooting method. There are s boundary values specified initially, and so s unknown values have to be estimated in order to start the process. We shall call these unknowns the parameters of the problem, $p_1,p_2,\ldots,p_s$ say. Let us take

$$\left.\begin{aligned} y_j(a) &= p_j, \quad j = k+1,\ldots,s,\\ y_{i_j}(b) &= p_j, \quad j = 1,2,\ldots,k, \end{aligned}\right\} \tag{1.4}$$

where $i_1, i_2,\ldots,i_k$ are all distinct and different from $i_{k+1},\ldots,i_s$ in (1.2). The forward integration starts with the given values $\alpha_1,\alpha_2,\ldots,\alpha_k$ and the estimates $p_{k+1},\ldots,p_s$ as initial conditions at $x = a$, and produces a solution $\underline{y}_a(x;\underline{p})$ which is a function of the parameters $\underline{p}$ (strictly of $p_{k+1},\ldots p_s$ only). Similarly the backward integration uses the values of β_i and the remaining values p_j as initial conditions at $x = b$, and produces a solution $\underline{y}_b(x;\underline{p})$. The two solutions are computed as far as some matching point $x = m$ in $[a,b]$. At this point they should be equal if the values of $\underline{p}$ are correct, so we get the following equations for determining $\underline{p}$

$$\underline{F}(\underline{p}) \equiv \underline{y}_a(m;\underline{p}) - \underline{y}_b(m;\underline{p}) = \underline{0}\;. \tag{1.5}$$

In general these are nonlinear equations in the s variables p_j, which have to be solved iteratively. The function $\underline{F}(\underline{p})$ is not known explicitly, but its value can be computed for any given values of $\underline{p}$ by numerical integration of (1.1), at least in principle. There may be various difficulties in the practical computation, of course, and these will be discussed later.

If the matching point m is chosen to be at a or b, we can reduce the number of parameters and the number of equations in (1.5). Suppose we take m at the point a, then we need only estimate k parameters at x=b, integrate backwards and match the first k variables of $\underline{y}_b(x;\underline{p})$ to the given boundary conditions at a. There are only k nonlinear equations to solve, and the forward integration is unnecessary. However, it is more convenient to discuss the method in general terms, and to assume that m is taken at a general point between a and b, giving a full set of s equations at the matching point. (This will also make it easier to handle wider classes of problems later on.) The algebra remains essentially the same if m = a or b, but some of the equations in (1.5) become trivial. The reason for not choosing the matching point at either a or b in all cases is to overcome certain practical problems which may arise in the integration of (1.1). If the system suffers from inherent instability in one direction, we can integrate in the reverse direction and choose the matching point at the appropriate end. This is not possible when the system is unstable in both directions, but the effect may be mitigated by moving the matching point towards the middle of the range. More serious cases of instability require special treatment, which will be considered in section 4.

If the differential equations (1.1) are linear, the forward and backward solutions $\underline{y}_a$, $\underline{y}_b$ are linear functions of the parameters $\underline{p}$. Then $\underline{F}(\underline{p})$ is also a linear function of $\underline{p}$ which may be written in the form

$$\underline{F}(\underline{p}) \equiv A\underline{p} - \underline{c} = \underline{0}, \tag{1.6}$$

where the matrix A and the vector $\underline{c}$ are not known initially. To find the solution of (1.6), we chose s+1 vectors $\underline{p}^{(t)}$, t = 0,1,...,s, such that the differences $\underline{p}^{(0)} - \underline{p}^{(j)}$, j = 1,2,...,s are linearly independent. We then calculate the corresponding values $\underline{F}^{(t)}$ of $\underline{F}(\underline{p})$ by integrating (1.1). The results may be written

$$\left.\begin{aligned} \underline{F}^{(0)} &= A\,\underline{p}^{(0)} - \underline{c}, \\ \underline{F}^{(j)} - \underline{F}^{(0)} &= A\,(\underline{p}^{(j)} - \underline{p}^{(0)}), \quad j = 1,2,\ldots,s, \end{aligned}\right\} \qquad (1.7)$$

from which we can determine A and $\underline{c}$, and hence solve the equations (1.6) for $\underline{p}$. However, it is not necessary to use a special procedure for the linear case, because the general iterative method to be described later will provide a solution quite simply (in one step).

In the case of nonlinear differential equations, we have a nonlinear system to solve in (1.5). The problem has three stages, finding an efficient method for computing $\underline{F}(\underline{p})$, selection of a suitable iterative process for solving (1.5), and choosing a good starting approximation for $\underline{p}$. To compute $\underline{F}(\underline{p})$ it is often satisfactory to use a simple integration method, e.g. of Runge-Kutta type, because the result is not required to high accuracy, at least in the early stages of the iteration. But the method must include reliable error checks, to ensure that the results are sufficiently accurate for the iteration to succeed. The attraction of the Runge-Kutta method is its simplicity and flexibility, especially for short ranges of integration. When the range of integration is long, e.g. for infinite ranges where the asymptotic conditions are not achieved quickly, it is likely that a more efficient high-order integration method should be chosen. In our experience the differential equations arising in boundary-value problems do not usually exhibit stiffness when solved as initial-value problems and so we do not need a special algorithm for stiff equations such as Gear's method. However, Guderley (1973) has shown by a simple example that stiffness can be a problem when solving invariant imbedding equations (see section 5 and Chapter 17). The problem of finding a good starting approximation may be the most difficult step for a general system, and it will be discussed later. The basic method used for solving (1.5) is usually the Newton iteration or one of its variants. We will consider this in detail after giving some examples of boundary-value problems.

A general account of shooting methods for two-point boundary-value problems is given by Keller (1968, Ch. 2). He also discusses conditions which guarantee the existence of a solution, though these are somewhat limited in practical application.

2. SOME EXAMPLES OF BOUNDARY-VALUE PROBLEMS

We consider first a simple example of a two-point boundary-value problem,

$$y'' = \sinh y,$$

with (2.1)

$$y'(0) = 0, \quad y(d) = c.$$

This equation arises from a one-dimensional form of the elliptic equation,

$$\nabla^2 \emptyset = k \sinh (q\emptyset), \tag{2.2}$$

which is derived from the Poisson-Boltzmann equation. If we convert (2.1) to the standard form, we have

$$y_1' = y_2,$$
$$y_2' = \sinh y_1$$

with (2.3)

$$y_2(0) = 0, \; y_1(d) = c \, .$$

Because the basic problem is elliptic, we would expect the differential system in (2.3) to be somewhat unstable for step-by-step solution. Integrating forwards from x = 0 or backwards from x = d is equivalent to solving a Cauchy problem for the elliptic equation, and it is well known that this is an inherently unstable process. The solutions of (2.3) are exponential in character, and over a long range the increasing exponential term tends rapidly to dominate the computed solution. Consequently, if we integrate forwards from x = 0 with two independent sets of boundary conditions, we may find that we get essentially the same solution after proceeding a certain distance. (If the exponential term increases rapidly, the solution may also go outside the range of the computer.) Similar considerations apply to the integration backwards from x = d. The difficulty is not insuperable in this case, and for short or moderate ranges the problem is easily solved if we control the instability by matching in the middle. But the example illustrates a common source of trouble when we use shooting methods for boundary-value problems, which often have solutions of exponential type.

Another example is taken from boundary-layer theory. The following equations arise in the study of free-convection flow (Szewczyk, 1964).

$$\left.\begin{aligned} y''' + 3yy'' - 2y'^2 + z &= 0, \\ z'' + 3pyz' &= 0, \\ \text{with} \quad y(0) = y'(0) = 0,\ z(0) &= 1, \\ y'(\infty) = 0,\ z(\infty) &= 0. \end{aligned}\right\} \tag{2.4}$$

If the system is written in the standard first-order form, we obtain a system of five differential equations. The special feature of this problem is that two of the boundary conditions are given at infinity, and we have to ensure that the solution has the correct asymptotic behaviour. We start by estimating the value of x at which asymptotic conditions are reached, x=R say, then carry out the integration and matching as usual. The solution is fairly stable for forward integration, and we can match at the end-point x=R, though it is difficult to get convergence unless we have good starting approximations to the values of $y''(0)$ and $z'(0)$. Having found a solution, the position of R is checked by studying the behaviour of y and z as $x \to R$. The correct solution must be such that y is approximately constant and z' approximately zero over the last few steps of the range. If these conditions are not fulfilled, we shift the point R outwards and repeat the calculation, using the previously determined solution as a starting approximation. In more complicated problems it is worth investigating the behaviour at infinity in some detail, and perhaps using the asymptotic form as a boundary condition. This will be considered in Chapter 18.

A third example is the case of Mathieu's equation with periodic boundary conditions

$$\left.\begin{aligned} y'' + (\alpha - 2q\cos 2x)\,y &= 0, \\ \text{with} \quad y(0) = y(2\pi),\ y'(0) &= y'(2\pi). \end{aligned}\right\} \tag{2.5}$$

It is known that a periodic solution exists for certain values of α

(Ince 1926). The differential equation is linear and it can be solved by fairly simple methods, provided α is not too large. If we formulate it for solution by the shooting method, we take the values of $y(0)$ and $y'(0)$ as parameters, noting that the first is essentially a scaling factor, because the differential equation is homogeneous. We compute a forward solution $\underline{y}_a(x;\underline{p})$ and obtain the matching condition at 2π directly from the periodic property, giving an equation to solve for $\underline{p}$ as usual. However, the value of $y(0)$ is indeterminate, because the parameters can be reduced to a single quantity, the ratio $y'(0)/y(0)$. So this case requires slightly different treatment from the standard problem, though it is not difficult to solve.

3. SOLUTION OF THE MATCHING EQUATIONS

We return now to the problem of solving the nonlinear system of equations (1.5). The best-known method for such systems is the Newton iteration, which takes the form

$$\underline{p}^{(t+1)} = \underline{p}^{(t)} - [J(\underline{p}^{(t)})]^{-1} \underline{F}(\underline{p}^{(t)}), \tag{3.1}$$

where $J(\underline{p})$ is the Jacobian matrix $[\partial F_i/\partial p_j]$, evaluated for parameters $\underline{p}$. We start with a suitable approximation $\underline{p}^{(0)}$, then calculate successive corrections by using (3.1) in the form

$$[J(\underline{p}^{(t)})]\ (\underline{p}^{(t+1)} - \underline{p}^{(t)}) = -\ \underline{F}(\underline{p}^{(t)}). \tag{3.2}$$

This represents a set of linear equations for the corrections to $\underline{p}^{(t)}$, which can be solved provided the Jacobian is non-singular.

To apply this method we have to calculate the s elements of the vector $\underline{F}(\underline{p}^{(t)})$ and the s^2 elements of the Jacobian matrix $J(\underline{p}^{(t)})$. Taking the general case where the matching point is in the middle of the range, we see from (1.4) and (1.5) that

$$\frac{\partial F_i}{\partial p_j} = \frac{\partial y_{ai}(m)}{\partial p_j}\ (j > k), \quad -\frac{\partial y_{bi}(m)}{\partial p_j}\ (j \leqslant k). \tag{3.3}$$

The computation of these derivatives is the lengthiest part of the process. They may be obtained in two ways, either by solving the variational equations for the system (1.1), or by approximating the derivatives using simple difference quotients.

Let us consider first the variational equations. We write

$$q_{ij} = \frac{\partial y_{ai}}{\partial p_j} \quad (j > k), \qquad \frac{\partial y_{bi}}{\partial p_j} \quad (j \leqslant k). \tag{3.4}$$

The quantities q_{ij} are functions of x and $\underline{p}$ defined in the range [a,m] for $j > k$, and in [m,b] for $j \leqslant k$. They satisfy the linearized equations

$$\frac{dq_{ij}}{dx} = \frac{\partial f_i}{\partial y_1} q_{1j} + \frac{\partial f_i}{\partial y_2} q_{2j} + \ldots + \frac{\partial f_i}{\partial y_s} q_{sj} \tag{3.5}$$

for i, j = 1,2,..., s, with boundary conditions

$$\left.\begin{aligned} q_{ij}(a) &= \delta_{ij}, \quad j > k, \\ q_{ij}(b) &= \delta_{ij}, \quad j \leqslant k. \end{aligned}\right\} \tag{3.6}$$

Thus to determine the q_{ij}, we have to integrate s - k sets of s linear differential equations forwards from a to m, and k sets of s equations backwards from b to m. The values of q_{ij} at x = m are then used in (3.3) to give the elements of the Jacobian.

For a simple boundary-value problem such as (2.1) it is quite easy to derive the variational equations (3.5), and to integrate them to obtain the quantities required for Newton's method. In cases where the variational equations are used, we should take account of their linear structure and solve directly for the q_{ij} by a stable implicit method. Riley, Bennett and McCormick (1967) discuss the use of high-order integration methods in this context. However, many systems are much more complex than (2.1), and finding the linearized equations is then a complicated algebraic process.

Instead of computing the q_{ij}, we can represent the derivatives by

$$\frac{\partial F_i}{\partial p_j} \simeq \frac{F_i(p_j + \delta p_j) - F_i(p_j)}{\delta p_j} \tag{3.7}$$

where the dependence of $\underline{F}(\underline{p})$ on the other components of $\underline{p}$ has been suppressed. To obtain the approximate Jacobian for use in (3.2), we have to integrate the original differential equation s+1 times, once with the current values of the parameters p_i, and once with each of the s parameters perturbed in turn. The number of integrations required is exactly the same as for the variational method, but in this case they all refer to the same system (1.1), which simplifies the programming. As with the variational equations, we only have to integrate as far as the matching point in each case.

A practical point which arises when we use finite differences is how to choose a suitable value for the perturbation δp_j. The equations (1.1) are integrated numerically, and consequently the values of $\underline{F}(\underline{p})$ are subject to the effects of rounding and truncation error. Assuming that the integration method has a reliable error control procedure, so that it can generally keep the local errors below a specified bound, we must take δp_j to be considerably greater than this bound. Otherwise the truncation errors in the integration will dominate in (3.7). However, there may still be serious cancellation in forming the difference quotient, particularly in problems which are somewhat unstable, when the effect of perturbing one of the parameters may be lost. In difficult cases the approximate Jacobian will have a number of very small elements, and it will appear to be nearly singular. If this happens we can try using the variational equations, but these will not necessarily give better results, because the equations (3.5) are likely to be unstable in the same way as the original system.

Convergence of the Newton method is discussed by Ortega and Rheinboldt (1970, §10.2). It is well known that the iteration has quadratic convergence near the solution (subject to certain regularity assumptions), but this is not usually a significant feature in practice. Most of the computing time needed to solve a boundary-value problem is spent in trying to find the neighbourhood of the solution. Until the iterates get close to the root, Newton's method may not be specially fast, and it is often better to modify it, as described below. In theory the method using differences has less then quadratic convergence near the solution (Ortega and Rheinboldt 1970, §11.2), but in practice its behaviour is very similar to that of Newton's method. Both methods fail in cases where the solution is not isolated (i.e. where it is close to a branch-point with respect to some parameter).

A number of variants of the Newton method have been suggested for boundary-value problems. One useful modification to the iteration (3.1) is to introduce a damping factor λ_t, so as to ensure that the norm of $\underline{F}(\underline{p})$ is reduced at each step. We write

$$\underline{p}^{(t+1)} = \underline{p}^{(t)} - \lambda_t \, [J(\underline{p}^{(t)})]^{-1} \, \underline{F}(\underline{p}^{(t)}), \tag{3.8}$$

where λ_t is a factor (usually $\leqslant 1$) chosen to ensure that $||\underline{F}(\underline{p}^{(t+1)})|| < ||\underline{F}(\underline{p}^{(t)})||$. Some safeguard such as this is essential if we are using Newton's method with a poor starting approximation, because the iterates can easily diverge if there is no check.

Another way of introducing a damping factor is given by the formula

$$\underline{p}^{(t+1)} = \underline{p}^{(t)} - [J(\underline{p}^{(t)}) + \mu_t I]^{-1} \, \underline{F}(\underline{p}^{(t)}). \tag{3.9}$$

Again the parameter μ_t can be used to ensure that $||\underline{F}(\underline{p})||$ is reduced at each step. It also provides a way of perturbing the problem in the case where the Jacobian $J(\underline{p})$ is singular or ill-conditioned for certain values of $\underline{p}$. The choice of λ_t in (3.8) or μ_t in (3.9) is often made empirically subject to the conditions mentioned, but these two methods can be very effective in practice. The convergence is superlinear under certain restrictions on the values of λ_t, μ_t (Ortega and Rheinboldt 1970, §11.2).

In cases where the chief problem is that of locating the root from a poor initial estimate, we can proceed more systematically by re-formulating it in terms of function minimization. If we take

$$S(\underline{p}) = ||\underline{F}(\underline{p})||^2 = \sum_{i=1}^{s} (F_i(\underline{p}))^2, \tag{3.10}$$

then the value of $\underline{p}$ which satisfies the equations $\underline{F}(\underline{p}) = \underline{0}$ also gives the minimum of the positive definite function $S(\underline{p})$. The gradient of $S(\underline{p})$ is $2[J(\underline{p})]^T \, \underline{F}(\underline{p})$, and so the steepest-descent method minimizing (3.10) has the form

$$\underline{p}^{(t+1)} = \underline{p}^{(t)} - 2\alpha_t J_t^T \underline{F}(\underline{p}^{(t)}), \tag{3.11}$$

where α_t is the step-length and $J_t = J(\underline{p}^{(t)})$. The value of α_t may be chosen by a line minimization technique. The matrix of second derivatives of $S(\underline{p})$ is given by

$$\frac{1}{2}\left[\frac{\partial^2 S}{\partial p_i \partial p_j}\right] = [J(\underline{p})]^T [J(\underline{p})] + \sum_{u=1}^{s} F_u(\underline{p}) \left[\frac{\partial^2 F_u}{\partial p_i \partial p_j}\right] \tag{3.12}$$

Neglecting the second term on the right (which may be expected to be small near the solution), we obtain the Gauss-Newton method for minimization.

$$\underline{p}^{(t+1)} = \underline{p}^{(t)} - [J_t^T J_t]^{-1} J_t^T \underline{F}(\underline{p}^{(t)}), \tag{3.13}$$

based on the quadratic approximation to $S(\underline{p})$. This reduces to Newton's method for solving $\underline{F}(\underline{p}) = \underline{0}$ when J is a square matrix.

The steepest-descent method may be the best way of starting the minimization, since it is guaranteed to reduce $S(\underline{p})$. However, its convergence is usually very slow as we approach the minimum. A more effective method is to use a combination of (3.11) and (3.13) in the form

$$\underline{p}^{(t+1)} = \underline{p}^{(t)} - [J_t^T J_t + \lambda_t I]^{-1} J_t^T \underline{F}(\underline{p}^{(t)}) \tag{3.14}$$

(Marquardt 1963). This is equivalent to a small steepest-descent step for large λ_t, and to a Gauss-Newton step for small λ_t.

This approach is the basis of some good recent algorithms for the solution of nonlinear equations, which can be used for solving boundary-value problems. However, it is not clear that sophisticated methods of solution make the best use of the special structure of the equations (1.5).

In the algorithm in the NAG Library (Mark 4, DO2ADA/F), we have always used the modified form of the Newton method (3.8) as follows:

Given an initial guess $\underline{p}^{(0)}$, set t=0 and perform the following steps:

(i) Calculate the matrix $J(\underline{p}^{(t)})$ and its triangular factors L_t, U_t by Gaussian elimination with partial pivoting so that

$$P_t J(\underline{p}^{(t)}) = L_t U_t \tag{3.15}$$

where P_t is a permutation matrix.

(ii) Calculate $\underline{z}$, $\underline{w}$ from

$$L_t \underline{z} = - P_t \underline{F}(\underline{p}^{(t)}), \quad U_t \underline{w} = \underline{z}.$$

(iii) For i = 1,2,... evaluate

$$\underline{p}_i = \underline{p}^{(t)} + \lambda_i \underline{w}, \quad \lambda_i = 1/2^{i-1},$$

and when

$$||\underline{F}(\underline{p}_i)||_2 < ||\underline{F}(\underline{p}^{(t)})||_2 \tag{3.16}$$

set $\underline{p}^{(t+1)} = \underline{p}_i$ and proceed to (iv).

(iv) If $||\underline{p}^{(t)} - \underline{p}^{(t+1)}||_\infty < e_1$, and $||\underline{F}(\underline{p}^{(t+1)})||_2 < e_2$ stop with $\underline{p}^{(t+1)}$ as the final value; if $||\underline{F}(\underline{p}_1)||_2^2 < 0.1||\underline{F}(\underline{p}^{(t)})||_2^2$ set $P_{t+1} = P_t$, $L_{t+1} = L_t$, $U_{t+1} = U_t$ and go to (iii) with t replaced by t+1; otherwise go to (i) with t replaced by t+1 (e_1, e_2 are suitably chosen error bounds).

This algorithm is a combination of Newton iteration and a minimization method, and given good enough estimated values $\underline{p}^{(0)}$ it has proved successful in practice. From poor starting values it often fails to terminate at stage (iii) or fails in calculating the factorisation (3.15)

A number of general methods for solving nonlinear equations are available in the NAG Library. We have tried the effect of substituting these algorithms for the one described above without obtaining any significant improvement in performance. In fact, in many cases the performance of the other methods was considerably worse, though one would expect that it could be improved by adapting the methods to the special form of the equations (1.5).

Deuflhard (1974) has considered in detail the problem of solving the matching equations (1.5). He uses an algorithm which is similar to the one just described but which differs in two important respects: the function $S(\underline{p})$ in (3.10) is modified, and a special algorithm is used when $J(\underline{p}^{(t)})$ is almost singular. He observes that most algorithms for solving $\underline{F}(\underline{p}) = 0$ minimise the function $S(\underline{p}) = ||\underline{F}(\underline{p})||_2$, and he suggests the modification $S(\underline{p}) = ||A\,\underline{F}(\underline{p})||_2$, where A is a matrix to be selected. In effect he uses a linear transformation to scale the vector $\underline{F}(\underline{p})$, and demonstrates that for this problem the best choice is $A = [J(\underline{p})]^{-1}$. He goes on to prove that the following method converges globally under mild conditions on $\underline{F}$ and J, the most severe of which is that $||[J(\underline{p})]^{-1}||$ is uniformly bounded for all $\underline{p}$ in the domain of interest (which includes $\underline{p}^{(0)}$ and the solution).

The t-th iterative step is to calculate

$$\underline{\Delta p}^{(t)} = -[J(\underline{p}^{(t)})]^{-1}\,\underline{F}(\underline{p}^{(t)}), \tag{3.17}$$

and to put

$$\underline{p}^{(t+1)} = \underline{p}^{(t)} + \lambda_t \underline{\Delta p}^{(t)}, \tag{3.18}$$

where λ_t is chosen to satisfy $0 < \lambda_{min} \leq \lambda_t \leq \lambda_{max} \leq 1$, and is such that

$$||[J(\underline{p}^{(t)})]^{-1}\,\underline{F}(\underline{p}^{(t+1)})||_2 < ||[J(\underline{p}^{(t)})]^{-1}\underline{F}(\underline{p}^{(t)})||_2. \tag{3.19}$$

Note the resemblance to the algorithm described earlier. The values λ_{min} and λ_{max} may not be easy to calculate, but Deuflhard remarks that they both approach unity as the solution is approached. In practice he takes $\lambda_{max} = 1$ and $\lambda_{min} = 1/8$ or $1/16$. (A similar practical convention is adopted in the NAG Library procedures.) Note that the condition (3.19) may be written

$$||\overline{\underline{\Delta p}}^{(t+1)}||_2 \leq ||\,\underline{\Delta p}^{(t)}||_2, \tag{3.20}$$

where

$$\overline{\underline{\Delta p}}^{(t+1)} = -[J(\underline{p}^{(t)})]^{-1}\underline{F}(\underline{p}^{(t+1)}).$$

and it is implemented in this form. There is an analogy here with iterative refinement for determining corrections to approximate solutions of linear systems of equations and this form of the Newton iteration is an equivalent process for nonlinear systems of equations.

Deuflhard suggests that when $J(\underline{p}^{(t)})$ is singular or almost singular we should use the pseudoinverse $[J(\underline{p}^{(t)})]^+$ in place of $[J(\underline{p}^{(t)})]^{-1}$. $J^+ = [J(\underline{p}^{(t)})]^+$ is defined as the matrix which satisfies

$$J^+ J J^+ = J^+, \; J J^+ J = J,$$

$$(J^+J)^T = J^+ J, \; (JJ^+)^T = J J^+,$$

He gives algorithms for computing the pseudoinverse when the rank s' of $J(\underline{p}^{(t)})$ is known, and discusses techniques for estimating s' using the well-known Businger and Golub (1965) algorithm, which can also be used to compute J^+. The accurate determination of rank is a very difficult problem, and Deuflhard remarks that numerical instability will be avoided as long as the rank is not overestimated. The smaller the value chosen for s' the smaller $||J^+||$ becomes, in general. We can find whether our choice of s' is too large by monitoring the convergence of the algorithm (3.17), (3.18) and (3.20). If at any stage no value of λ_t can be found such that (3.20) is satisfied, we reduce s' and try again. When the initial estimate $\underline{p}^{(0)}$ is known to be a poor approximation to the solution, Deuflhard suggests taking a small s' initially and increasing it as the iteration proceeds. Further details are given in his paper, which also includes some interesting numerical examples.

Let us turn our attention briefly to some of the other problems associated with shooting methods. The problem of determining a good initial vector $\underline{p}^{(0)}$ is often critical. In the cases where conventional mathematical techniques fail, we can adopt a variety of numerical strategies. We mentioned above Deuflhard's technique of reducing the estimated rank of the Jacobian initially in order to get started, and increasing it progressively as we approach the solution. His examples indicate that this approach can be very useful. Other techniques include (i) parallel shooting (see section 4 below), which may sometimes work with less accurate initial guesses, (ii) an invariant imbedding algorithm (see section 5 below) to obtain a low-accuracy solution, which then gives starting values for the main calculation and (iii) a continuation (homotopy) method as described in Chapter 18. Deuflhard uses continuation and rank reduction in some of his numerical examples.

Many practical problems are given on infinite ranges of integration and the equations may also have singularities. These problems will be discussed in detail in Chapter 18, but we observe here that in the case of infinite ranges, we often have asymptotic values for more than s variables, though only s boundary conditions are truly independent. In such a case we may consider minimising $||\underline{F}(\underline{p})||_2$ where $\underline{F}$ has s components and $\underline{p}$ has less than s. If the corresponding (rectangular) Jacobian $J(\underline{p})$ has full column rank, we can apply the Gauss-Newton method (3.13) directly. We have not found this strategy very successful, on the whole, but a report by Nachtscheim and Swigert (1965) suggests that it can be useful. It is likely that the major difficulty in using (3.13) for overdetermined systems is ill-conditioning of J^TJ, which is a well-known problem in the analogous linear least squares equations. The iteration

$$\underline{p}^{(t+1)} = \underline{p}^{(t)} - [J(\underline{p}^{(t)})]^+ \underline{F}(\underline{p}^{(t)})$$

is mathematically equivalent to (3.13), but if implemented carefully will give fewer problems with numerical stability. It seems therefore that Deuflhard's ideas may also be useful in this situation.

4. PARALLEL SHOOTING METHODS

For problems where the integration of (1.1) across [a,b] is difficult or impossible because of instability, we consider subdividing the range [a,b] in some way. The most commonly used method of this type is parallel shooting (Keller 1968). Consider a partition

$$a = x_o < x_1 < x_2 < \dots < x_n = b, \tag{4.1}$$

where the spacing of the points $x_1, x_2, \dots x_{n-1}$ is chosen to reflect the difficulty of integrating (1.1) over different parts of the range. That is we choose the points x_i close together where (1.1) is particularly difficult to integrate; this is discussed further below. We consider first the general boundary conditions

$$A\underline{y}(a) + B\underline{y}(b) = \underline{\alpha} \tag{4.2}$$

in place of (1.2). Define

$$\begin{gathered} \Delta_j = x_j - x_{j-1}, \quad t = (x - x_{j-1})/\Delta_j, \\ \underline{y}_j(t) = \underline{y}(x_{j-1} + t\Delta_j), \quad \underline{f}_j(t,\underline{y}_j) \equiv \Delta_j \underline{f}(x_{j-1} + t\Delta_j, \ \underline{y}_j) \end{gathered} \tag{4.3}$$

With this change of variables, equation (1.1) with boundary conditions (4.2) becomes

$$\frac{d\underline{y}_j}{dt} = \underline{f}_j(t,\underline{y}_j(t)), \quad 0<t<1, \quad j = 1,2,\dots n, \tag{4.4}$$

and

$$A\ \underline{y}_1(0) + B\ \underline{y}_n(1) = \underline{\alpha}\,. \tag{4.5}$$

At each interior point we require continuity of the solution, that is

$$\underline{y}_{j+1}(0) = \underline{y}_j(1), \quad j = 1,2,\ldots, n-1, \tag{4.6}$$

Combining equations (4.4) - (4.6), we obtain the following equations to solve

$$\frac{d\underline{z}}{dt} = \underline{g}(t,\underline{z}), \qquad 0<t<1, \tag{4.7}$$

with boundary conditions

$$P\underline{z}(0) + Q\underline{z}(1) = \underline{\gamma}, \tag{4.8}$$

where

$$\underline{z}(t) = \begin{pmatrix} \underline{y}_1(t) \\ \underline{y}_2(t) \\ \cdot \\ \cdot \\ \cdot \\ \underline{y}_n(t) \end{pmatrix}, \quad \underline{g}(t,\underline{z}) = \begin{pmatrix} \underline{f}_1(t,\underline{y}_1) \\ \underline{f}_2(t,\underline{y}_2) \\ \cdot \\ \cdot \\ \cdot \\ \underline{f}_n(t,\underline{y}_n) \end{pmatrix}, \quad \underline{\gamma} = \begin{pmatrix} \underline{\alpha} \\ \underline{0} \\ \cdot \\ \cdot \\ \cdot \\ \underline{0} \end{pmatrix}, \tag{4.9}$$

$$P = \begin{pmatrix} A & & & & & \\ & I & & & 0 & \\ & & I & & & \\ & & & \cdot & & \\ & 0 & & & \cdot & \\ & & & & & \cdot \\ & & & & & I \end{pmatrix}, \quad Q = \begin{pmatrix} 0 & 0 & \cdot & \cdot & \cdot & B \\ -I & 0 & \cdot & \cdot & \cdot & 0 \\ & -I & \cdot & & & \cdot \\ & & \cdot & \cdot & & \cdot \\ & & & \cdot & \cdot & \cdot \\ & 0 & & & \cdot & 0 \quad 0 \\ & & & & -I & 0 \end{pmatrix}.$$

We can apply the shooting method to the problem (4.7), (4.8) exactly as in section 1. That is we attempt to choose a vector $\underline{p}$ such that the solution $\underline{u}(\underline{p};t)$ of the initial value problem

$$\frac{d\underline{u}}{dt} = \underline{g}(t,\underline{u}), \quad \underline{u}(\underline{p};0) = \underline{p} \tag{4.10}$$

satisfies the nonlinear equations

$$\underline{F}(\underline{p}) \equiv P\underline{p} + Q\,\underline{u}(\underline{p};1) - \underline{\gamma} = \underline{0}\,. \tag{4.11}$$

All our earlier remarks concerning the solution of this system of nonlinear equations still apply. If we use a Newton method, the basic iteration has the form

$$\underline{p}^{(t+1)} = \underline{p}^{(t)} - [J(\underline{p}^{(t)})]^{-1}\ \underline{F}(\underline{p}^{(t)}), \tag{4.12}$$

and we can calculate the Jacobian $J(\underline{p})$ either by numerical differentiation or by solving variational equations. The variational equations for (4.10) can be obtained by differentiating (4.10) with respect to $\underline{p}$, giving

$$\frac{dU}{dt} = \frac{\partial \underline{g}}{\partial \underline{u}}(t,\underline{u}(\underline{p};t))U, \quad U(\underline{p};0) = I, \tag{4.13}$$

where U is a matrix of order ns, and (4.12) can be written

$$\underline{p}^{(t+1)} = \underline{p}^{(t)} - (P+QU(\underline{p}^{(t)};1))^{-1}\ \underline{F}(\underline{p}^{(t)}). \tag{4.14}$$

Note that in practice we integrate (4.10) as n separate systems of order s, and (4.13) as n separate systems of order $s \times s$.

The related problems of the condition of equation (4.11) and the location of the points x_i have been discussed by several authors. Keller (1968) shows that if (A+B) is nonsingular then (P+Q) is nonsingular. He goes on to show that $P+Q\ U(\underline{p}^{(t)};1)$ is nonsingular if

$$||(P+Q)^{-1}Q||_\infty ||U(\underline{p}^{(t)};1) - I||_\infty < 1, \tag{4.15}$$

and observes that

$$||U(\underline{p}^{(t)};1) - I||_\infty \leqslant \max_{1\leqslant j\leqslant n} [\exp(\int_0^1 k(x_{j-1} + t\Delta_j)dt) - 1], \tag{4.16}$$

where k(x) is any function such that

$$\left\| \frac{\partial \underline{f}}{\partial \underline{y}} (x,\underline{y}) \right\|_{\infty} \leq k(x), \quad a < x < b. \tag{4.17}$$

Hence the closer the points x_i, especially in the parts of the range $[a,b]$ where $k(x)$ is large, the better the chance of $(P + QU)$ being nonsingular and, indeed, the better the condition of $(P+QU)$. In fact we might choose the points x_i so as to minimise the right hand side of inequality (4.16); this is the philosophy of George and Gunderson (1972) who consider the linear system $\underline{y}' = A(x)\underline{y} + \underline{f}(x)$ in great detail. We describe their analysis in Chapter 17 but remark here that it may also be possible to apply it to nonlinear problems. In general the requirement that $(P+Q)$ is nonsingular is unlikely to be satisfied; for example, for the separated boundary conditions (1.2), $(A+B)$ is nonsingular only if $i_k = k$. However all that is really necessary for the solution of (1.1) and (4.2) is that the matrix $[A,B]$ has rank s, a fact which George and Gunderson use. Osborne (1974) has also considered the choice of the points x_i so as to make $(P+QU)$ well-conditioned, demonstrating that if $\Delta_j = x_j - x_{j-1}$ is small for all j we have the required property. He recommends choosing the number of points x_i as a function of the size of $\| \partial\underline{f}/\partial\underline{y} \|$ and this idea is clearly related to minimising the right hand side of (4.16). We believe that a simple practical algorithm for choosing the points x_i would be an important step in the development of parallel shooting methods.

Keller (1968) gives an algorithm for solving the system of equations

$$(P + QU)\underline{\Delta p}^{(t)} = -\underline{F}(\underline{p}^{(t)}) \tag{4.18}$$

arising from (4.14), which exploits the block structure of the matrix $P+QU$. We must be particularly careful in using the block structure, because the fact that $P+QU$ is well-conditioned only implies that a conventional Gaussian elimination technique applied to $P+QU$ will be safe. Any algorithm based on block elimination must be analysed independently. Osborne (1974) and Deuflhard (1974) also discuss this problem, as does Weiss (1974) though his matrices arise from a different technique for boundary-value problems. Deuflhard extends his method described earlier to parallel shooting for the general boundary conditions (4.2).

In the case of separated boundary conditions of the form (1.2), we solve only for the unknown boundary values and the solution at the points x_i. In this case the ns × ns matrix P+QU has the form

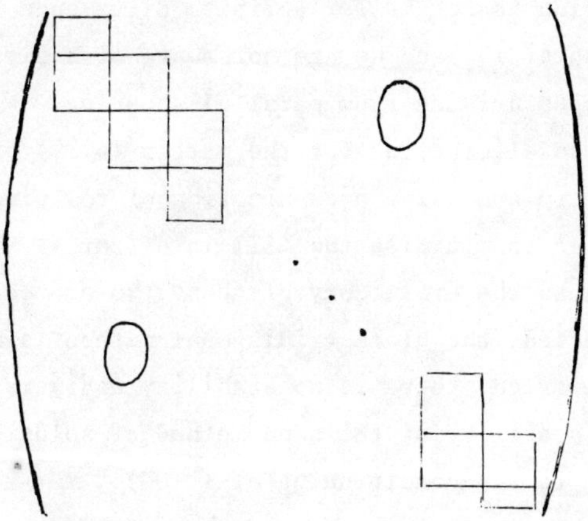

(4.19)

where only elements inside the blocks are non-zero and where each block is s×s except for the first which is k×s and the last which is (s-k)×s. By appending blocks of zeros we can write this matrix in block tridiagonal form

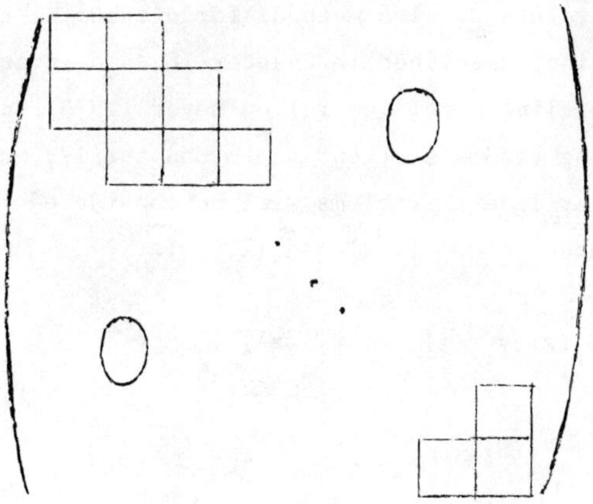

(4.20)

where each block is an $s \times s$ matrix. Varah (1973) and Keller (1974) discuss the stability of a block-tridiagonal elimination procedure on (4.20) using pivoting within the diagonal blocks but not between blocks. Their discussion is for Keller's finite difference approximation (see Chapter 15) and we are not aware of a similar analysis for the matrices arising from parallel shooting. Weiss (1974) and Varah (1972) discuss elimination for the matrix (4.19) preserving only its banded structure and using a combination of row pivoting and column pivoting designed to minimise the fill-in of zeros. Varah (1972) and Keller (1974) discuss the efficiency of these two procedures and conclude that, as expected, the block tridiagonal method is more efficient However, as we have remarked, there is no stability analysis for this algorithm whereas the stability of the band method of solution is justified by appealing to the conditioning of (P+QU).

For an example of the parallel shooting method applied in practice see Cebici and Keller (1971).

5. INVARIANT IMBEDDING FOR OBTAINING ESTIMATES TO BOUNDARY VALUES.

It has been emphasised above that the main difficulty in applying an iterative technique for solving (1.5) often lies in estimating the unknown boundary values $\underline{p}$. Two methods for overcoming the problem are continuation, described in Chapter 18, and invariant imbedding, which we outline here. We follow Meyer (1973) in our discussion and draw the reader's attention to Chapter 17, where invariant imbedding for linear problems will be considered in more detail. Using the subscripts of (1.2) and (1.4) let

$$\begin{aligned} \underline{v}(x) &= [y_{i_1}(x), y_{i_2}(x), \ldots, y_{i_k}(x)]^T, \\ \underline{u}(x) &= [y_{i_{k+1}}(x)\ y_{i_{k+2}}(x), \ldots, y_{i_s}(x)]^T \end{aligned} \qquad (5.1)$$

where we presume that a good initial guess for $\underline{v}(b)$ is required to start the shooting method from $x = b$. Now write the problem (1.1), (1.2) as

$$\underline{u}' = \underline{f}_1(x,\underline{u},\underline{v}),\ \underline{v}' = \underline{f}_2(x,\underline{u},\underline{v}), \tag{5.2}$$

$$\underline{u}(a) = A\underline{v}(a) + \underline{g}_1, \tag{5.3}$$

$$\underline{u}(b) = \underline{g}_2, \tag{5.4}$$

where A is an $(s - k)\times k$ matrix and $\underline{g}_1$ and $\underline{g}_2$ are vectors. The invariant imbedding algorithm as described by Meyer (1973) is:

Step 1 Integrate over [a,b] the initial value problem

$$\left.\begin{aligned} \frac{\partial \underline{u}}{\partial x}(t,\underline{v}) + \frac{\partial \underline{u}}{\partial \underline{v}}\underline{f}_2(x,\underline{u},\underline{v}) &= \underline{f}_1(x,\underline{u},\underline{v}), \\ \underline{u}(a,\underline{v}) &= A\,\underline{v}(a) + \underline{g}_1\,. \end{aligned}\right\} \tag{5.5}$$

Step 2 Solve the system of nonlinear equations

$$\underline{u}(b,\underline{v}) = \underline{g}_2 \tag{5.6}$$

for $\hat{\underline{v}}(b)$.

Step 3 Solve the initial value problem

$$\left.\begin{aligned} \underline{u}' &= \underline{f}_1(x, \underline{u}, \underline{v}), \quad \underline{u}(b) = \underline{g}_2 \\ \underline{v}' &= \underline{f}_2(x, \underline{u}, \underline{v}), \quad \underline{v}(b) = \hat{\underline{v}}(b) \end{aligned}\right\} \tag{5.7}$$

backwards to $x = a$.

To solve equation (5.6), we must be able to obtain $\underline{u}(b,\underline{v})$ for any $\underline{v}$ and to solve $\underline{u}(b,\underline{v}) = \underline{g}_2$. This and the fact that we must solve the first-order system of hyperbolic equations (5.5) for $\underline{u}$ over a large set of vectors $\underline{v}$, implies that the dimension k of $\underline{v}$ must be small. Similarly the dimension of $\underline{u}$ must be small or the computational cost will be high. The equation (5.5) can be solved by the Lax-Wendroff method, or any of the other stable explicit methods described by Meyer. If the solutions of (5.5) and (5.6) are obtained sufficiently accurately then the solution of the ordinary differential equations will yield an accurate solution to (5.2) - (5.4). In any case the value $\hat{\underline{v}}(b)$ obtained from (5.6) should be a good estimate for the correct value, and hence it can be used to supply components for $\underline{p}$ in the shooting method. Meyer (1973) gives several interesting numerical examples of small nonlinear systems of the form (1.1) and (1.2) which have been solved by invariant imbedding, and the values he obtains could be used as starting values to obtain more accurate solutions from a shooting or parallel shooting method.

17

LINEAR DIFFERENTIAL EQUATIONS AND DIFFERENTIAL EIGENPROBLEMS

1. INTRODUCTION

In this chapter, we turn from discussing the nonlinear boundary value problem

$$\underline{y}' = \underline{f}(x,\underline{y}), \quad B\underline{y}(a) + C\underline{y}(b) = \underline{\gamma}, \quad a \leqslant x \leqslant b, \tag{1.1}$$

to the simpler linear problem

$$\underline{y}' = A(x)\underline{y} + \underline{f}(x), \quad B\underline{y}(a) + C\underline{y}(b) = \underline{\gamma} \tag{1.2}$$

where $A(x)$, B and C are square matrices of order s. We observe that there is a considerable literature, see for example Coddington and Levinson (1955), on the theory of this topic. Indeed it is possible to contruct an explicit solution for constant matrices $A(x) = A$, which may even be computable for small problems. Many methods have been devised which depend on particular properties of equation (1.2), and we describe some of the more useful here.

We then turn to the linear differential eigenproblem

$$\underline{y}' = A(x,\lambda)\underline{y}, \quad B\underline{y}(a) + C\underline{y}(b) = \underline{0} \tag{1.3}$$

and discuss how the techniques devised for (1.2) may be used in this case. Finally we briefly discuss by an example the problem of tracing the eigenvalues λ of

$$\underline{y}' = A(x,\lambda,\mu)\underline{y}, \quad B\underline{y}(a) + C\underline{y}(b) = \underline{0} \tag{1.4}$$

as the parameter μ varies.

2. METHODS FOR LINEAR BOUNDARY VALUE PROBLEMS

We consider the problem (1.2) except that to simplify the discussion we use separated boundary conditions. That is, we consider the problem of finding the vector $\underline{y}(x) = [y_1(x), y_2(x), \ldots, y_s(x)]^T$ which satisfies

$$\underline{y}' = A(x)\underline{y} + \underline{f}(x), \tag{2.1}$$

$$y_i(a) = \alpha_i, \; i=1,2, \ldots, k, \tag{2.2}$$

$$y_{i_j}(b) = \beta_j, \; j = k+1, k+2, \ldots, s, \tag{2.3}$$

where $1 \leqslant i_{k+1} < i_{k+2} < \ldots \; i_s \leqslant s$. We can, of course, use any of the algorithms described earlier for nonlinear equations, ignoring the linearity of the problem (2.1) - (2.3), but we cannot expect to obtain a particularly efficient algorithm in this way. We shall mainly be concerned with special forms of the shooting method for (2.1) - (2.3), but first we mention that the finite difference method, maybe with extrapolation (Keller 1969), can be applied as described in Chapter 15. In this method we must solve a banded system of linear equations for each step-length used. To achieve a required accuracy we may be faced with solving large systems of linear equations and this will inevitably involve using relatively large amounts of computer storage, especially if extrapolation is used. However, finite difference methods do tend to be more stable for elliptic-type problems than the shooting methods described below. In contrast shooting methods mainly involve small systems of linear equations but the computational cost of forming these equations is much greater. As far as we know, no thorough comparison has been made between finite difference and shooting algorithms. The expansion methods to be described in Chapter 19 can also lead to competitive algorithms for solving linear differential systems. Varah (1973), Russell and Varah (1974) and Russell (1975) discuss finite difference methods with extrapolation and expansion methods using collocation on piecewise polynomial splines and give an interesting comparison of these methods.

The shooting method described in Chapter 16 is mathematically equivalent to the following method. Solve (s+1) initial value problems

$$Z'(x) = A(X)Z(x), \quad Z(a) = I \tag{2.4}$$

$$\underline{z}'(x) = A(x)\underline{z}(x) + \underline{f}(x), \quad \underline{z}(a) = \underline{0} \tag{2.5}$$

where Z(x) is a square matrix of order s. (Note that solving equation (2.4) is equivalent to solving s initial value problems, one for each column of Z.) Then select the vector $\underline{d}$ so that

$$\underline{y}(x) = \underline{z}(x) + Z(x)\underline{d} \tag{2.6}$$

satisfies the boundary conditions (2.2) and (2.3), and using the superposition property we see that the resulting vector $\underline{y}(x)$ is the solution of (2.1) - (2.3). Of course when we solve the initial value problems (2.4), (2.5) numerically, we only obtain an approximation to $\underline{y}(x)$ in (2.6). More efficiently, we can obtain the solution by solving only s-k initial value problems in place of (2.4) since we have only s-k boundary conditions at x = b. For example, we might solve the initial value problems

$$\underline{z}'(x) = A(x)\underline{z}(x) + \underline{f}(x), \quad \underline{z}(a) = [c_1, c_2, \ldots, c_k, 0, \ldots, 0]^T \tag{2.7}$$

$$Z'(x) = A(x)Z(x), \; Z(a) = E \tag{2.8}$$

where Z(x) and E are s×(s-k) matrices and E has columns $\underline{e}_{k+i}$, i = 1,2,...,s-k, ($\underline{e}_j$ is the j-th column of the unit matrix of order s). We form the combination $\underline{y}(x) = \underline{z}(x) + Z(x)\underline{d}$ which satisfies the boundary conditions (2.3). Many variants of this method have been suggested in the literature, for example the well-known method of adjoints is mathematically equivalent. We refer the interested reader to the review in Roberts and Shipman (1972b).

In practice, the simple methods described above often fail. There are at least two distinct reasons why failure may occur. If $A(x)$ has at least one positive eigenvalue for all $x \in [a,b]$ the equations (2.4), (2.5), (2.7) and (2.8) may be inherently unstable and the integration will then break down before $x = b$ is reached. If all the eigenvalues of $A(x)$ are positive for all $x \in [a,b]$ we should integrate from $x = b$ to $x = a$. If $A(x)$ has widely separated eigenvalues at least on a large subinterval $[c,b]$ we may be able to perform the integrations but the computed matrix $Z(b)$ will have almost linearly dependent columns since each column will be approximately proportional to the eigenvector of $A(x)$ corresponding to the dominant eigenvalue (or, if there is a group of dominating eigenvalues, to a combination of the corresponding eigenvectors).

To avoid these difficulties, in particular the problem of inherent instability, we may use a parallel shooting method as described in Chapter 16. In fact, as we shall see, this method can also assist in reducing the problem of linear dependence. Consider the following linear form of the parallel shooting method (Keller 1968) described in Chapter 16 for nonlinear problems. Choose a partition $a = x_0 < x_1 < \ldots < x_n = b$, then with $\Delta_j = x_j - x_{j-1}$, $t = (x - x_{j-1})/\Delta_j$, define

$$\underline{z}_j'(t) = A_j(t)\underline{z}_j(t) + \underline{f}_j(t), \quad j = 1,2,\ldots,n, \tag{2.9}$$

for $t \in [0,1]$ where $A_j(t) = \Delta_j A(x)$, $\underline{f}_j(t) = \Delta_j \underline{f}(x)$. Integrating

$$\underline{z}_j'(t) = A_j(t)\underline{z}_j(t) + \underline{f}_j(t), \quad \underline{z}_j(0) = r_j, \; j = 1,2,\ldots,n, \tag{2.10}$$

and

$$Z_j'(t) = A_j(t)Z_j(t), \quad Z_j(0) = I, \quad j = 1,2,\ldots,n \tag{2.11}$$

from $t = 0$ to $t = 1$, we solve the problem (1.2) by computing the solution of the linear system of equations

$$(P + QZ(1))\Delta\underline{r} = -(P\underline{r} + Q\underline{z}(1) - \underline{\alpha}) \tag{2.12}$$

where

$$P = \begin{pmatrix} A & & & \\ & I & & 0 \\ & & \ddots & \\ 0 & & & I \end{pmatrix} \qquad Q = \begin{pmatrix} 0 & & & B \\ -I & 0 & & \\ & -I & \ddots & \\ & & -I & 0 \end{pmatrix},$$

$$Z(t) = \begin{pmatrix} Z_1(t) & & 0 \\ & Z_2(t) & \\ 0 & & \ddots \; Z_n(t) \end{pmatrix}, \quad \underline{r} = \begin{pmatrix} \underline{r}_1 \\ \underline{r}_2 \\ \vdots \\ \underline{r}_n \end{pmatrix}, \quad \underline{z}(t) = \begin{pmatrix} \underline{z}_1(t) \\ \underline{z}_2(t) \\ \vdots \\ \underline{z}_n(t) \end{pmatrix} \quad \underline{\alpha} = \begin{pmatrix} \underline{\gamma} \\ \underline{0} \\ \\ \underline{0} \end{pmatrix}.$$

If $\underline{\Delta r} = (\underline{\Delta r}_1, \underline{\Delta r}_2, \ldots, \underline{\Delta r}_n)^T$, then $\underline{r}_i + \underline{\Delta r}_i$ is the correct value of $\underline{y}(x_{i-1})$; a sensible choice of $\underline{r}$ would be $\underline{r} = 0$. For the boundary conditions (2.2) and (2.3), the equation (2.12) can be simplified to a banded system of equations, see Chapter 16 for a discussion.

Of course in a numerical solution of (2.10), (2.11) and (2.12) we obtain only an approximate solution to equation (1.2), but it is not the numerical solution which is usually the problem but the inherent instability of the differential equation. George and Gunderson (1972) analyse the condition of linear boundary value problems with regard to solving the equation (2.12). Their analysis can be used to determine the number and position of the points x_i in the partition of $[a,b]$ so that the equations (2.12) remain reasonably well-conditioned. If we define the condition number

$$M[Z(t)] = \|Z(t)\| \; \|Z(t)^{-1}\|$$

for, say, the Euclidean norm, then George and Gunderson show that, if $0 < \ell < 1$ and

$$\int_{x_{j-1}}^{x_j} \|A(s)\| \, ds \leqslant \log_e \left[\frac{\ell}{\|(P+Q)^{-1}\| \; \|P\|} \right] + 1, \tag{2.13}$$

then

$$M[P + QZ(1)] \leqslant M[P + Q] \; M[Z(1)] \left(\frac{1+\ell}{1-\ell}\right). \tag{2.14}$$

Defining the logarithmic norm

$$\mu[A] = \lim_{h\to 0} \; (\|I + hA\| - 1)/h,$$

George and Gunderson show that

$$M[Z(1)] \leqslant \exp\{\max_j \int_{x_{j-1}}^{x_j} \mu[A(s)]\,ds + \max_j \int_{x_{j-1}}^{x_j} \mu[-A(s)]\,ds\}.$$

Hence the right hand side of inequality (2.14) can be estimated as long as $\max \Delta_j$ is small enough so that (2.13) is satisfied. Obviously the bound on $M[Z(1)]$ and hence on $M[P + QZ(1)]$ can be made as small as necessary by making $\max_j \Delta_j$ as small as necessary. However as $\Delta_j \to 0$, the size of the system (2.12) increases and computational difficulties could arise here. Guderley (1973) discusses a parallel shooting algorithm which involves integrating the adjoint equation in addition to (2.11). He also makes many of the points discussed by George and Gunderson including discussing the condition of equations (2.12).

An alternative but related technique for solving inherently unstable problems is the orthonormalisation technique of Conte (1966). To solve problem (2.1)-(2.3), we solve a sequence of initial value problems. At the first stage we solve the initial value problems (2.7), (2.8) for $x > a$ noting that at $x = a$, $\underline{z}(a)$ and the columns of $Z(a)$ are orthogonal. We integrate towards $x = b$ whilst the columns $\underline{Z}_i(x)$ of $Z(x)$ remain 'sufficiently linearly independent.' We use the criterion that

$$\min_{ij} \cos^{-1} \left| \frac{\underline{Z}_i(x)^T \underline{Z}_j(x)}{\{\underline{Z}_i(x)^T \underline{Z}_i(x)\}^{\frac{1}{2}} \{\underline{Z}_j(x)^T \underline{Z}_j(x)\}^{\frac{1}{2}}} \right| \geqslant \alpha \qquad (2.15)$$

as a check that the columns of $Z(x)$ are linearly independent. If this criterion is not satisfied at $x = x_1$ say, we reorthogonalise the columns $\underline{Z}_i(x)$ there. If α is taken to be $\frac{\pi}{2}$, this test will almost always fail and we will have to reorthogonalise at each step of the numerical integration, whereas if $\alpha = 0$, the test will never fail and Conte's method reduces to the simple superposition method described earlier. Conte suggests using a small value for α so as to avoid performing an unnecessarily large number of reorthogonalisations. He also suggests as a simpler check the replacement of (2.15) by

$$\max_j |\underline{Z}_j(x)^T \underline{Z}_j(x)| < M$$

for some constant $M>0$. This check may not be very reliable as it depends too much on the growth characteristics of the matrix $Z(x)$. At the reorthogonalisation point x_1, we orthonormalise the columns of $Z(x_1)$ by choosing P_1 to give the orthogonal matrix

$$Z^1(x_1) = Z(x_1)P_1,$$

and then we orthogonalise $\underline{z}(x_1)$ with respect to the columns of $Z^1(x_1)$ by choosing the vector $\underline{w}_1$ so that

$$\underline{z}^1(x_1) = \underline{z}(x_1) - Z^1(x_1)\underline{w}_1 \qquad (2.16)$$

is orthogonal to the columns of $Z^1(x)$. We then proceed by integrating (2.7) and (2.8) from x_1 with initial conditions $Z(x_1) = Z^1(x_1)$ and $\underline{z}(x_1) = \underline{z}^1(x_1)$ until we reach a point x_2 where we must reorthogonalise again, using a matrix P_2 and vector $\underline{w}_2$, and so on until we reach $x = b$. The orthogonalisation (2.15) and (2.16) is performed by Conte (1966) using a Gram-Schmidt process. For large systems of equations it would be better to use the Modified Gram-Schmidt process described by Björck (1967). To obtain the final solution assume that there have been m orthogonalisations on the range $[a,b]$ and that we have stored sufficient information to calculate P_i, $\underline{w}_i$, $i = 1,2,\ldots, m$. We choose the vector $\underline{\beta}_m$ in the solution $\underline{y}(x) = \underline{z}(x) + Z(x)\underline{\beta}_m$ to satisfy the boundary conditions (2.3). We can then recover the solution at earlier orthogonalisation points by generating the vectors $\underline{\beta}_i$ from

$$\underline{\beta}_{i-1} = P_i[\underline{\beta}_i - \underline{w}_i], \quad i = m, m-1, \ldots, 1. \qquad (2.17)$$

The value of $\underline{y}(x)$ between x_i and x_{i+1} (with $x_0 = a$, $x_{m+1} = b$) is given by

$$\underline{y}(x) = \underline{z}(x) + Z(x)\underline{\beta}_i \qquad (2.18)$$

where $\underline{z}(x)$ and $Z(x)$ are the integrated solutions on $[x_i, x_{i+1}]$. This algorithm with its great flexibility is justifiably a favourite among superposition methods for solving linear systems of equations. A thorough recent analysis has been given by Scott and Watts (1975).

An alternative approach to the difficulties arising from inherent instability when solving boundary value problems by shooting techniques is to modify the initial value problems being solved. This is our main reason for considering invariant imbedding techniques. Here we follow the development in Meyer (1973). Guderley (1973), Scott (1973a) and the chapters by Meyer and Scott in Aziz (1975) are also interesting contributions to this topic. This invariant imbedding method includes as special cases some of the projection method techniques of Guderley and Nicolai (1966) and the factorisation techniques of Babuska, Prager and Vitasek (1966).

Consider the linear boundary value problem

$$\begin{aligned} \underline{u}' &= A_1(x)\underline{u} + A_2(x)\underline{v} + \underline{f}_1(x) \\ \underline{v}' &= A_3(x)\underline{u} + A_4(x)\underline{v} + \underline{f}_2(x), \end{aligned} \tag{2.19}$$

$$\underline{u}(a) = F_1\underline{v}(a) + \underline{a}_1, \quad F_2\underline{u}(b) + F_3\underline{v}(b) = \underline{a}_2 \tag{2.20}$$

where $\underline{u}$, $\underline{f}_1$ and $\underline{a}_1$ are s_1- and $\underline{v}$, $\underline{f}_2$ and $\underline{a}_2$ are s_2-component vectors respectively with $s_1 + s_2 = s$, and A_1, A_2, A_3, A_4, F_1, F_2 and F_3 are matrices of appropriate order. Given the problem (2.1) - (2.3), we can always write it in the form (2.19) and (2.20) though the choice of which components of $\underline{y}$ in (2.1) are to be assigned to $\underline{u}$ and which to $\underline{v}$ is not necessarily clear. In some cases, for example when the boundary conditions (2.20) can be written $\underline{u}(a) = \underline{a}_1$, $\underline{v}(b) = \underline{a}_2$, the division is clear, whereas in others it may depend implicitly on the properties required of the differential equations below. Meyer (1973) discusses these points in some detail and we shall return to some of them later. The algorithm given by Meyer for solving (2.19) and (2.20) follows:

Step 1 Integrate to x = b, the Riccati system

$$U' = A_2(x) + A_1(x)U - UA_4(x) - UA_3(x)U, \; U(a) = F_1, \tag{2.21}$$

integrate to x = b, the linear system

$$\underline{w}' = \left[A_1(x) - U(x)A_3(x)\right]\underline{w} - U(x)\underline{f}_2(x) + \underline{f}_1(x), \; \underline{w}(a) = \underline{a}_1, \tag{2.22}$$

Step 2 Find the solution $\hat{\underline{v}}(b)$ of the linear equations

$$[F_2U(b) + F_3]\hat{\underline{v}}(b) = a_2 - F_2\underline{w}(b) \tag{2.23}$$

Step 3 Integrate the linear system

$$\underline{v}' = [A_3(x)U(x) + A_4(x)]\ \underline{v} + A_3(x)\underline{w}(x) + \underline{f}_2(x),\ \underline{v}(b) = \hat{\underline{v}}(b), \tag{2.24}$$

backward over $[a,b]$.

The solution to the boundary value problem (2.19), (2.20) is then

$$\underline{y}(x) = \begin{bmatrix} \underline{u}(x) \\ \underline{v}(x) \end{bmatrix} \qquad \text{where } \underline{u}(x) = U(x)\underline{v}(x) + \underline{w}(x).$$

One difficulty with this algorithm is the backward integration on Step 3 where values of $U(x)$ and $\underline{w}(x)$ are required for backward integration of $\underline{v}(x)$. We may not have sufficient computer storage to keep the appropriate values from Step 1. If this is the case, we can integrate the full set of equations (2.19) with initial conditions $\underline{u}(b) = U(b)\underline{v}(b) + \underline{w}(b)$ and $\underline{v}(b)$. However, one of the aims of the invariant imbedding algorithm is to avoid integrating the full set of equations because of their inherent instability (Meyer 1973). Other techniques for avoiding this full integration have been suggested; see Scott's contribution in Aziz (1975).

Another difficulty may arise when solving the Riccati equation (2.21) where the true solution can in some situations become unbounded. Of itself, this need not lead to any difficulty since the numerical solution will become unbounded at the same point and good initial value programs will terminate before this happens. As long as $U(x)$ is a square matrix (as it will be if $s_1 = s_2$) we can form $U^{-1}(x)$ and then integrate a related Riccati equation for $U^{-1}(x)$, recovering the original solution using well-known relations after reaching $x = b$.

To see an advantage of the invariant imbedding algorithm consider the problem

$$u'' - u = 0, \quad u(0) = 1, \quad u'(b) = e^{-b}. \tag{2.25}$$

Since the general solution of this differential equation is $C_1e^x + C_2e^{-x}$ we obtain almost linearly dependent solutions for the superposition method with any starting value $u'(0)$, when b is large enough. Writing the differential equation (2.25) as the system

$$u' = v, \quad v' = u$$

in the invariant imbedding algorithm we must solve the Riccati equation

$$U' = 1 - U^2, \quad U(0) = 0$$

whose exact solution is $U = \tanh(x)$ which is bounded by unity for positive x. The solutions of (2.22) and (2.24) in this case are exponentially decreasing and we never encounter the problems of the superposition method.

If the range of integration is $[a,\infty]$ then the conventional approach using any of the superposition methods described above is to solve the problem for a sequence of intervals $[a,b_i]$ with $b_i \to \infty$ as $i \to \infty$. We apply the final conditions at the point b_i in each case, stopping when consistent results have been obtained. It is possible to devise an invariant imbedding algorithm for this problem too. Details are given in Alspaugh (1974) which includes some interesting numerical examples.

See Scott (1974) for an extensive bibliography for invariant imbedding and related topics. Many of the problems mentioned above have been discussed at length in the literature but a more general treatment is outside the scope of this chapter.

3. LINEAR DIFFERENTIAL EIGENPROBLEMS

Consider the following simplified form of problem (1.3)

$$\underline{y}' = A(x,\lambda)\underline{y} \tag{3.1}$$

$$y_1(a) = y_2(a) = \ldots = y_k(a) = y_{i_1}(b) = y_{i_2}(b) = \ldots = y_{i_k}(b) = 0, \tag{3.2}$$

where $\underline{y}(x) = [y_1(x), y_2(x), \ldots, y_s(x)]^T$, $s = 2k$, and $1 \leq i_1 < i_2 < \ldots < i_k \leq s$. If we use a finite difference method as in Chapter 15 or an expansion method as in Chapter 19, we obtain an algebraic eigenproblem

$$B(\lambda)\underline{z} = 0 \tag{3.3}$$

where $B(\lambda)$ is a matrix depending (nonlinearly) on λ and, in the case of finite difference approximation, $\underline{z}$ is a vector of approximations to $\underline{y}(x)$ at the mesh points $x = x_i$. In the case of an expansion method $\underline{z}$ is a vector of coefficients in an expansion for $\underline{y}$. If $A(x,\lambda)$ in (3.1) can be written $A(x,\lambda) = C(x)+\lambda D(x)$, then for sensible approximations $B(\lambda) = E + \lambda F$ where E and F are independent of λ. In general, for finite difference approximations $B(\lambda)$ in equation (3.3) is too large a matrix for the compuation of all its eigenvalues to be feasible. The usual approach is to solve the equation $\det(B(\lambda)) = 0$ for the particular values of λ of interest. Methods which operate directly with $B(\lambda)$ such as inverse iteration for the smallest eigenvalue λ (when B is linear in λ) or simultaneous iteration for a group of eigenvalues are often useful (Wilkinson 1965). For an expansion method, if the expansion is chosen carefully, the matrix $B(\lambda)$ may be small enough so that algorithms which compute all the eigenvalues of $B(\lambda)$ (when it is linear) can be used.

To use the superposition methods described above we may proceed as follows. For any value λ, we solve the initial value problem

$$Y' = A(x,\lambda)Y, \quad Y(a,\lambda) = I \tag{3.4}$$

to obtain the value $Y(b,\lambda)$. Using the superposition property of solutions of linear systems of differential equations we obtain

$$\underline{y}(b) = Y(b,\lambda)\underline{y}(a) \tag{3.5}$$

for any initial values $\underline{y}(a)$. Now applying the boundary conditions (3.2), we see that if we partition

$$Y(b,\lambda) = \left[G,H\right] \tag{3.6}$$

where G and H are s x k matrices then

$$\underline{y}(b) = H\underline{z}(a) \tag{3.7}$$

where $\underline{z}(a) = [y_{k+1}(a), y_{k+2}(a),\ldots,y_s(a)]^T$. If the rows $i_1, i_2,\ldots,i_k$ of H are taken as the rows of the matrix K then $K\underline{z}(a) = 0$ from (3.2). Hence λ is an eigenvalue if $\det(K) = 0$. Any superposition technique described earlier can be used to obtain $Y(b,\lambda)$, or preferably just H in (3.7). Gersting and Jankowski (1972) survey methods of this type applied to the Orr-Sommerfeld equation

$$\{(D^2-\alpha^2)^2 - i\alpha R[(\bar{u}-c)(D^2-\alpha^2)-D^2\bar{u}]\}\phi = 0 \tag{3.8}$$
$$D\phi(0) = D^3\phi(0) = \phi(1) = D\phi(1) = 0$$

where D denotes differentiation, α is the wave number, R is the Reynolds number and $c = c_r + ic_i$ is the eigenvalue, and the function $\bar{u}$ is known. They prefer Conte's orthonormalisation method for this problem. A full description of the use of Conte's method is given by Davey (1973) who adds a few improvements. The use of superposition methods for solving eigenvalue problems is fraught with the same difficulties as in solving linear boundary value problems.

In an attempt to overcome these difficulties, Prüfer transformations and Riccati equations have been used by some writers. For example, consider the Sturm-Liouville problem

$$(p(x)\psi'(x))' + (\lambda r(x) - q(x))\psi(x) = 0, \tag{3.9}$$
$$\psi(a) \cos \alpha - p(a)\psi'(a) \sin \alpha = 0,$$
$$\psi(b) \cos \beta - p(b)\psi'(b) \sin \beta = 0.$$

We define the phase function $\theta(x)$ satisfying the equations

$$\psi(x) = \rho(x) \sin \theta(x), \quad p(x)\psi'(x) = \rho(x) \cos \theta(x). \tag{3.10}$$

Then it can be shown that

$$\theta'(x) = \frac{1}{p(x)} \cos^2\theta(x) + (\lambda r(x) - q(x)) \sin^2\theta(x) \tag{3.11}$$

with boundary conditions

$$\theta(a) = \alpha, \quad \theta(b) = \beta + n\pi \tag{3.12}$$

where $0 \leqslant \alpha < \pi$, $0 < \beta \leqslant \pi$. By setting $n = 1,2,\ldots$, we obtain the eigenvalues $\lambda_1, \lambda_2, \ldots$ in turn by integrating (3.11) with initial value $\theta(a) = \alpha$ and solving the nonlinear equation $\theta(b) = \beta + n\pi$ for λ. This method has been discussed for the problem (3.9) where $p(x) > 0$, $q(x) > 0$ in Godart (1965), and extensions to singular problems $(p(a)=0)$, problems on infinite ranges $(a = -b = -\infty)$ and fourth order differential equations have been described in Bailey (1966), Banks and Kurowski (1968), Soop (1968), and Banks and Kurowski (1973).

Scott, Shampine and Wing (1969) describe an invariant imbedding algorithm for the problem (3.9) and indicate how this method relates to the method of Godart (1966). Scott (1973b) extends this invariant imbedding algorithm to systems of the form (3.1) and remarks that the phase function technique described above does not itself extend easily. Jankowski, Takeuchi and Gersting (1972) describe the invariant imbedding algorithm in the context of the Orr-Sommerfeld problem (3.8) whilst Alspaugh, Kagiwada and Kalaba (1970) apply the algorithm to a special fourth order problem arising in structural mechanics.

All the shooting methods we have described involve finding a zero of a determinantal function of the parameter λ. Recalling the Orr-Sommerfeld equation (3.8), we observe that any of the values α, R, or c can be treated as the eigenvalue, if the others are known. It is often the case that the stability curve $c_i = 0$ in the $\alpha - R$ plane is required. If we know one point on this curve then we can trace the curve by varying, say, α slowly and solving the equation $c_i = 0$ as a function of R. The previously calculated value R will be a good approximation to its correct value provided α has not changed much. We solve a (nonlinear) determinantal equation for c for each pair of values α, R and use the value c_i thus obtained in the search for the zero of $c_i = 0$ (i.e. another nonlinear equation with argument R.) Any standard nonlinear equation-solving technique should be successful as long as the steps in α are small. This example is typical of many stability curve problems.

18

THE METHOD OF CONTINUATION AND GENERALIZED BOUNDARY VALUE PROBLEMS

1. INTRODUCTION

In this chapter we consider first the continuation method, which extends the range of problems that can be solved by shooting techniques. Then we discuss some generalized types of boundary value problem, where the parameters do not occur as simple boundary conditions. Such problems can be investigated by a more general form of the matching program described earlier.

2. PROBLEM OF STARTING CONDITIONS

The use of the shooting method for boundary-value problems leads in general to a system of nonlinear equations for the unknown parameters of the problem. In the simple case discussed in Chapter 16, the parameters are boundary values of the variables y_i. We assume as before that we have s ordinary differential equations of the form

$$\underline{y}' = \underline{f}(x,\underline{y}) \tag{2.1}$$

in the range $[a,b]$, with s boundary values of y_i specified, some at x=a and some at x=b. The missing boundary values are taken as the parameters p_j, $j = 1,2,\ldots,s$. The equation for $\underline{p}$ takes the form

$$\underline{F}(\underline{p}) \equiv \underline{y}_a(m;\underline{p}) - \underline{y}_b(m;p) = 0 \tag{2.2}$$

where $\underline{y}_a$ and $\underline{y}_b$ are the forward and backward solutions from x=a and x=b respectively, and m is the matching point.

In principle the equation $\underline{F}(\underline{p}) = 0$ can be solved by Newton's method, provided the Jacobian $J(\underline{p})$ is non-singular and not rapidly varying in the neighbourhood of the solution. This implies that the solution of

(2.2), and hence of the boundary-value problem, is isolated. The case of branch points, where two neighbouring solutions coalesce, is of interest in some practical applications, but it will not be considered here (see Kubicek and Hlavacek, 1974). However, even for isolated solutions there is no guarantee that Newton's method will converge unless the starting value of $\underline{p}$ is a good approximation to the exact solution. The usual conditions for convergence from a general starting point (Section 3) are extremely stringent, and unlikely to be fulfilled in practice. This is true for any nonlinear algebraic equation $\underline{F}(\underline{p}) = 0$, but in the case of boundary-value problems there are some special difficulties. Such problems are often derived from differential systems of elliptic type in two or three dimensions, and consequently they are unstable for step-by-step solution. Thus the computed values of $\underline{F}(\underline{p})$ may be inaccurately determined or even discontinuous with respect to $\underline{p}$, for values of $\underline{p}$ which are fairly close to the true solution. Whether this is due to numerical instability or to analytical properties of the solution makes little difference in practice. In either case the shooting method is very sensitive to small changes in the initial estimate of $\underline{p}$, and it may be impossible to make progress with any simple form of Newton iteration, or with minimization techniques as described in Chapter 16.

An example of this behaviour is given by Roberts and Shipman (1972). The problem

$$\left.\begin{aligned} y'' &= k \sinh ky, \\ y(0) &= 0,\ y(1) = 1, \end{aligned}\right\} \tag{2.3}$$

has a well-behaved solution in $[0,1]$ for k = 5, but this solution has a pole not far from x=1. For the shooting method, we integrate forward from x=0 with an estimated value of y'(0). If the estimate is too large, the pole moves closer to the origin, and it may fall within the range of integration. Any numerical method of solution for the differential equation will then give overflow, and so $\underline{F}(\underline{p})$ cannot be computed.

In such cases it is not possible to solve (2.2) by using an algorithm for nonlinear equations directly. Instead we have to take account of the structure of the boundary-value problem, and try to find a path of approach to the solution which avoids unstable regions. The basic idea of the continuation method is to find a problem of similar structure to the given problem which can be solved without difficulty, and to perturb it in stages into the original problem. The method depends on the assumption that the solution will vary continuously as the problem is perturbed, and so each stage will provide a good starting approximation for the solution at the next stage. There is great freedom of choice in the type of perturbation, in the problem used as a starting point, and in the method of tracing the solution. We shall describe various approaches which have proved useful, without exhausting all the possibilities.

In general algebraic terms, we want to find a function $\underline{G}(\underline{p},t)$, depending on the s parameters $\underline{p}$ and an additional variable t, chosen so that the equation

$$\underline{G}(\underline{p},0) = 0 \qquad (2.4)$$

is easy to solve, and so that $\underline{G}(\underline{p},1) \equiv \underline{F}(\underline{p})$. We then solve the series of equations

$$\underline{G}(\underline{p},t) = 0,\ t = 0,\ t_1,\ t_2,\ \ldots, \qquad (2.5)$$

starting with t=0 and increasing it steadily. When t=1 we have solved the original problem. Comparing with (2.2), $\underline{G}(\underline{p},t)$ corresponds to the matching function for the boundary-value problem, and so we require a method of introducing the variable t into the original differential system.

One simple way of perturbing the boundary-value problem is by varying the range of integration. If the system is very unstable over $[a,b]$, but fairly easy to solve over a shorter range $[a,c]$, we take a sequence of intermediate ranges

$$a \leqslant x \leqslant tb + (1-t)c, \qquad (2.6)$$

starting with the easy case t=0. This is particularly useful for an infinite range, where b is finally taken as the point where asymptotic conditions are reached. It may be impossible to integrate directly as far as x=b until we have a good approximation to the unknown boundary values at x=a. So we solve initially over a short range, without expecting to reach true asymptotic conditions, and use the solution as the starting approximation for a slightly longer range, and so on. In (2.6) we take $t=0$ initially, and then increase t in steps until the solution has the correct asymptotic behaviour.

A more general method is to perturb the differential equation (2.1) directly in some way; for example we can write

$$\underline{y}' = t\underline{f}(x,\underline{y}) + (1-t)\underline{g}(x,\underline{y}), \tag{2.7}$$

where $\underline{g}(x,\underline{y})$ is chosen so that the problem is easy to solve with $t=0$. The original problem is recovered when $t=1$. However, (2.7) is too general a form to give much insight into the problem of continuation. It is obvious that difficulties will arise if we choose an unsuitable function for $\underline{g}(x,\underline{y})$; for example, $\underline{g} = 0$ gives $\underline{y}$ = constant when $t = 0$, and we may find that we cannot satisfy all the boundary conditions. To overcome this difficulty we can introduce the parameter t into the boundary conditions as well, but we are constraining the continuation method to start from a somewhat arbitrary point in function space. The method is most likely to succeed if we can choose the function $\underline{g}$ so that the initial solution has a similar form to the required solution.

The theoretical basis of continuation is not yet fully developed, in particular there is very little theory to guide us in choosing the t-path effectively. However, the method is an important advance in handling difficult nonlinear problems, although it still depends to a large extent on experiment and judgement. In the next section we shall discuss some convergence results, and then go on to consider techniques for implementing the method.

3. CONDITIONS FOR CONVERGENCE

A basic result about the region of convergence for Newton's method is given by Kantorovich (1964, p. 708). We will quote a slightly different form of the theorem. For the equation $\underline{F}(\underline{p}) = 0$, let $J(\underline{p})$ be the Jacobian matrix $|\partial F_i/\partial p_j|$, and let $\underline{p}_0$ be the starting point of the iteration. Then the Newton method converges to a solution in the region $R : \| \underline{p} - \underline{p}_0 \| \leqslant r$ if the following conditions are satisfied:

$$\left.\begin{array}{ll}\text{(i)} & ||J(\underline{q}_1) - J(\underline{q}_2)|| \leqslant c||\underline{q}_1 - \underline{q}_2||, \text{ for } q_1, q_2 \text{ in R},\\ \text{(ii)} & c \leqslant 1/(2Kd), \text{ where } K = ||J(\underline{p}_0)^{-1}||,\\ & \qquad\qquad d = ||J(\underline{p}_0)^{-1}\,\underline{F}(\underline{p}_0)||,\\ & r \geqslant 1/(cK).\end{array}\right\} \tag{3.1}$$

The limitation of these conditions is that they are sufficient but not necessary for convergence, and they are too stringent for practical application. They are not usually satisfied until convergence of the iteration is well-established. The essential requirement for Newton's method is that the Jacobian should be non-singular, which is ensured by the above conditions for a certain region about $\underline{p}_0$. However, it is only necessary that $J(\underline{p})$ should be non-singular in the neighbourhood of the actual path followed by the iterates, and in other directions its behaviour is not important.

In the case of the continuation method, the equation (2.5) must have a solution for all t in $[0,1]$. Let $S(\underline{p},t)$ be the Jacobian matrix for $\underline{G}(\underline{p},t)$ with respect to $\underline{p}$, i.e. $S(\underline{p},t) = [\partial G_i/\partial p_j]$; then we must have $S(\underline{p},t)$ non-singular for any t in $[0,1]$, where $\underline{p}$ is the solution of $\underline{G}(\underline{p},t) = 0$ for the same value of t. We assume that the solution of the problem $\underline{G}(\underline{p},0) = 0$ is available. To ensure the theoretical feasibility of continuation along the t-curve, we require the Newton method to converge for a series of problems

$$\underline{G}(\underline{p}, k\Delta t) = 0,\ k = 1, 2, \ldots, N, \tag{3.2}$$

where Δt is a finite increment in t and $N\,\Delta t = 1$, using the previous solution as a starting point for each problem. This is discussed by Avila (1974), who establishes the following sufficient conditions: that $\underline{S}(\underline{p},t)^{-1}$ should exist and be bounded at all points of the t-curve, and that

$$||S(\underline{q}_1,t) - S(\underline{q}_2,t)|| \leqslant c_1||\underline{q}_1 - \underline{q}_2|| \tag{3.3}$$

for all $\underline{q}_1$, $\underline{q}_2$, in an open region containing the t-curve. We do not require analogues of conditions (ii) and (iii) of (3.1), because we can take Δt sufficiently small (though finite) to ensure that they are satisfied. (In practice it is not necessary to take Δt constant; it can vary according to the behaviour of the iteration.)

The significant conclusion from the convergence results is that the path of continuation must not go through any branch-points or other singularities of the perturbed boundary-value problem. To take a simple example, suppose we want to solve the problem

$$\left.\begin{aligned} &y'' + \alpha^2 y = g(x), \\ \text{with}\quad &y(0),\ y(1) \text{ given.} \end{aligned}\right\} \tag{3.4}$$

If for some reason we decide to proceed by continuation on the length of the range, we might start with the range $[0, 0.5]$ say and gradually extend it. The problem has eigensolutions for homogeneous boundary conditions when the right-hand end-point is at π/α, $2\pi/\alpha$, etc. and for these points the Jacobian will be singular. Taking $\alpha=4$, the first value occurs between 0.5 and 1, and the conditions for continuation do not hold. However, it might be possible to compute the solution in practice, because we work with a discrete set of points $x_1, x_2, x_3, \ldots,$ between 0.5 and 1, and the Jacobian will be singular only if one of the x_i is precisely at the point $\frac{\pi}{\alpha}$. But the problem would be ill-conditioned for values of x_i close to this point.

Further discussion of the convergence of the continuation method may be found in Ortega and Rheinboldt (1970, §10.4). We have considered above only the use of pure Newton iteration for tracing the solution along the t-curve. In practice we could use various modifications of the method and obtain a more robust algorithm, but with small steps in t it is not necessary to introduce much complexity.

4. DIFFERENTIAL EQUATIONS FOR THE T-CURVE

Some of the early papers on continuation were published by Davidenko (see Rall, 1968, for a summary), and he takes a more analytical approach to the problem. In his description of the method he obtains the differential equations with respect to t which are satisfied by the family of solutions along the continuation curve. The basic equation $\underline{F}(\underline{p}) = 0$ is imbedded in the set of equations

$$\underline{G}(\underline{p},t) = 0, \quad 0 \leq t \leq 1, \tag{4.1}$$

and for any particular value of t we regard (4.1) as an equation for determining $\underline{p}$; hence $\underline{p}$ is a function of t, $\underline{p} = \underline{p}(t)$. Differentiating (4.1) with respect to t gives

$$\sum_{j=1}^{s} \frac{\partial G_i}{\partial p_j} \frac{\partial p_j}{\partial t} = - \frac{\partial G_i}{\partial t}, \; i = 1,2,\ldots,s, \tag{4.2}$$

or $S(p,t) \dfrac{d\underline{p}}{dt} = - \dfrac{\partial \underline{G}}{\partial t}$.

This is a system of a first-order differential equations for $\underline{p}(t)$, to be solved as an initial-value problem in the interval $[0,1]$ with the starting condition $\underline{p}(0) = \underline{p}_0$ say, where $\underline{p}_0$ satisfies the equation $\underline{G}(\underline{p},0) = 0$.

The coefficients $\frac{\partial G_i}{\partial p_j}$ are functions of the current values of t and $\underline{p}(t)$, and they may be obtained by solving the variational equations as described in Chapter 16. Suppose we are using continuation in the form (2.7), and the parameters p_j are simple boundary values of the variable y_i. By differentiating (2.7) with respect to p_j we get

$$\frac{d}{dx}\left(\frac{\partial y_i}{\partial p_j}\right) = \sum_{k=1}^{s} \left\{ t \frac{\partial f_i}{\partial y_k} + (1-t) \frac{\partial g_i}{\partial y_k} \right\} \frac{\partial y_k}{\partial p_j} \tag{4.3}$$

for i = 1,2, ..., s, a linear system of equations for $\frac{\partial y_i}{\partial p_j}$ with boundary conditions of the form $\frac{\partial y_i}{\partial p_j} = \delta_{ij}$ corresponding to the given conditions at x=a or x=b. By solving these equations we can evaluate the matrix $S(\underline{p},t)$. (Alternatively the derivative with respect to p_j may be approximated by differences as in Chapter 16.) To evaluate the right-hand side of (4.2), we have from (2.7)

$$\frac{d}{dx}\left(\frac{\partial y_i}{\partial t}\right) = \sum_{k=1}^{s} \left\{ t \frac{\partial f_i}{\partial y_k} + (1-t) \frac{\partial g_i}{\partial y_k} \right\} \frac{\partial y_k}{\partial t} + (f_i - g_i), \tag{4.4}$$

giving a nonhomogeneous linear system to solve. The boundary values of $\frac{\partial y_i}{\partial t}$ corresponding to the specified boundary conditions for y_i are zero. We note that equations (4.3) and (4.4) have the same form, except for the nonhomogeneous term.

Since the process of determining the coefficients in (4.2) is rather complicated, it is clear that the method used for integrating (4.2) should not require a large number of function evaluations. A simple approach is to use Euler's method

$$\underline{p}(t_k+\Delta t_k) = \underline{p}(t_k) - \Delta t_k \left\{ S(\underline{p},t)^{-1} \times \frac{\partial G}{\partial t} \right\}_{t=t_k} \tag{4.5}$$

However, this gives results of rather low accuracy, and we must either take very small steps Δt_k or (preferably) correct the values obtained after every few steps by re-solving the original nonlinear system (2.7).

Examples of the continuation method in the form just described are given by Wasserstrom (1973).

5. PRACTICAL APPLICATION

We have considered two types of continuation process, (i) extension of the range of integration, and (ii) perturbation of the given differential equations. These two methods are not essentially distinct; we can convert the first type into the second by a transformation of the independent variable. But in practical work it is simpler to take the equations as they are given, and to use either (i) or (ii) as appropriate. We assume that multiple shooting will be used wherever necessary, i.e. where the solutions are too unstable for step-by-step solution even with good starting values (see Chapter 16). It is quite easy to combine, say, range extension and multiple shooting, and we leave the details to the reader.

In the second type of continuation process, the perturbation of the differential equation may be carried out in many different ways, of which (2.7) is just one example. The problem of Roberts and Shipman (1972) quoted earlier may be solved for large k by introducing the variable t in the following manner (cf. eqn. 2.3))

$$\left. \begin{aligned} y'' &= \tfrac{1}{2}\, k\, (e^{tky} - e^{-ky}), \\ y(0) &= 0,\ y(1) = 1. \end{aligned} \right\} \tag{5.1}$$

For small t the solution is fairly stable and it can be found without difficulty. As t increases to 1, (5.1) approaches the original problem. Although the system becomes very unstable, it is possible to trace the solution as $t \to 1$ by using multiple shooting.

Returning to the form (2.7) for the purpose of discussion, the continuation process is likely to be most effective when the solution for t = 0, satisfying

$$\underline{y}' = \underline{g}(x,\underline{y}), \tag{5.2}$$

bears some relation to the solution of the original problem. If the two solutions are entirely different in character, there is less likelihood of being able to perturb one problem smoothly into the other. It is difficult to give a practical prescription for choosing $\underline{g}(x,\underline{y})$ in general. One approach is to linearize $\underline{f}(x,\underline{y})$ and take $\underline{g}(x,\underline{y})$ to be equal to the linear terms, i.e.

$$\underline{g}(x,\underline{y}) = \underline{f}(x,\underline{y}^{(0)}) + \left(\frac{\partial f_i}{\partial y_j}\right)_{\underline{y}^{(0)}} (\underline{y} - \underline{y}^{(0)}), \tag{5.3}$$

where $\underline{y}^{(0)}$ is the current best estimate of the solution. This method and some variants are discussed by Roberts and Shipman (1973). However, if the linearized equations can be solved easily, it may be more profitable to pursue the Newton iteration directly in function space, without bringing in continuation. In problems which are strongly nonlinear, a perturbed form such as (5.1), which involves t nonlinearly, is probably better than (2.7).

As regards the computation of the solution as t varies, the important point is that the intermediate solutions for $0 \leqslant t \leqslant 1$ are generally of no interest in themselves; all that is wanted is a quick way of getting from t = 0 to t = 1. The method of Section 4, using the explicit differential equations along the t-curve, seems to be unnecessarily complicated. If the shooting method is used throughout, all that is needed to pass from t_k to $t_{k+1} = t_k + \Delta t_k$ is a good estimate of the values of the parameters $\underline{p}(t)$ at t_{k+1}. If Δt_k is small, $\underline{p}(t_k)$ is often a good enough approximation, but a more sophisticated form of extrapolation may be used to improve efficiency. This point is discussed by Rheinboldt (1975), actually in the context of finite-element methods, but his remarks can be applied to other problems. He suggests that $\underline{p}(t_{k+1})$ should be estimated by a Lagrangian extrapolation formula, using previous values of $\underline{p}(t)$ at t_k, t_{k+1},... (Clearly this is only possible after the first step.) The step-length Δt_k is chosen from the condition that the estimated error of the extrapolated value of $\underline{p}$ should be less than the 'radius of attraction' of the iteration process at t_{k+1}. This procedure has a sound theoretical foundation, but in practice it is not easy to estimate the radius of attraction very accurately. More experience is needed with such methods.

6. GENERAL PARAMETER STRUCTURE

In discussing the shooting method we have considered mainly the simplest form of boundary-value problem, where the unknown parameters are the values of y_i at the end-points of the range. The method of shooting and matching can be applied to a wider variety of problems, where the unknown quantities are not necessarily boundary values, but may be parameters occurring in the equation (2.1) or in the boundary conditions, or the range of integration. The general structure of the problem remains the same; there are a number of parameters p_j which are to be determined from a nonlinear equation of the form

$$\underline{F}(\underline{p}) \equiv \underline{y}_a(m;\underline{p}) - \underline{y}_b(m;\underline{p}) = 0. \tag{6.1}$$

This represents the matching condition at $x = m$ for two solutions $\underline{y}_a$ and $\underline{y}_b$, obtained by integrating the system (2.1) from $x = a$ and $x = b$ respectively (cf. Chapter 16). The system (6.1) consists of s matching equation for the s components of $\underline{y}$, and we would expect that in general s parameters could be determined. (In exceptional cases there may be fewer than s parameters, so that (6.1) is an over-determined system. We discuss this possibility later.)

To illustrate the generalized boundary-value problem, we consider some simple examples. The system

$$\left.\begin{aligned} &y'' + ky = f(x), \\ \text{with}\quad &y(0) = \alpha_1,\ y'(0) = \alpha_2, \\ &y(b) = \beta_1,\ y'(b) = \beta_2, \end{aligned}\right\} \tag{6.2}$$

is second-order, and the differential equation may be written in the form of (2.1). In the standard type of boundary-value problem, all the quantities in (6.2) will be known except α_2 and β_2 say. Because the problem is linear, the missing conditions can easily be determined by shooting or by some other method. A more general type of problem arises when the four quantities α_1, α_2, β_1, β_2 are given, but the coefficient k and the position of the end-point b are unknown. We can then regard k and b as the parameters of the problem; we estimate their values and integrate the equations, and then solve a matching equation of the form (6.1) to find them accurately.

Another example is given by

$$\left.\begin{aligned} &2yy'' + y'^2 = \tfrac{8}{9}(7y + 2x^2), \\ \text{with}\quad &y(0) = 0,\ y(1) = 2. \end{aligned}\right\} \tag{6.3}$$

This has the structure of a simple boundary-value problem, but we note that there is a singularity at the origin, where the first derivative is unbounded. The parameters would normally be taken as $y'(0)$ and $y'(1)$, but we cannot integrate the equation from $x = 0$ because of the singularity. It is quite easy to show that

$$y \sim p_1 x^{2/3} + \text{higher powers} \tag{6.4}$$

for small x, where p_1 is an unknown coefficient. So we take as parameters p_1 and $p_2 = y'(1)$, and try to determine them from the matching procedure. For this equation we start the integration not at the end-point $x = 0$, but a short distance away, at $x = h$ say. The value of h must be small enough to allow us to neglect the higher powers in (6.4), but not so small as to make the integration difficult. At the point h we can express both y and y' in terms of p_1, and proceed with the shooting method as usual. It is advisable to repeat the calculation with a slightly different value of h, to check that h has been chosen suitably.

A third type of problem arises when we have asymptotic boundary conditions, for example in the following equations for free-convection boundary layer flow

$$\left.\begin{aligned} y''' + 3yy'' - 2y'^2 + z &= 0, \\ z'' + 3\alpha yz' &= 0, \\ \text{with} \qquad y(0) = y'(0) = 0,\ z(0) &= 1, \\ y'(x) \to 0,\ z(x) \to 0 \text{ as } x &\to \infty. \end{aligned}\right\} \tag{6.5}$$

This is a fifth-order system, and it is clear that we can take the two missing boundary conditions at $x = 0$ as two of the unknown parameters. As $x \to \infty$, $y(x)$ tends to a constant value, and this provides a third parameter. The other two missing boundary conditions are the values of $y''(x)$ and $z'(x)$ at infinity, which are not really unknown, because they must be zero when the asymptotic conditions are reached. So it appears that we have three parameters and five matching conditions, giving an over-determined problem. In a case like this, it is possible to solve the equations (6.1) in the least-squares sense, and if the problem is properly formulated, the solution will given an exact match for all components. But in our experience it is often more satisfactory to include two extra parameters and determine them in the usual way, although their values are known to be zero. It is better still to investigate the problem analytically, and to obtain the leading terms of the asymptotic expansions for y and z. This gives the parameters in a different form, as coefficients or exponents of the local solution at infinity, and it ensures that we have the proper relations between y, y', and y'', and z, z'. In many cases it has the additional advantage of reducing the effective length of the range; the asymptotic form will be accurate long before y' and z are zero, as in (6.5).

To illustrate, we consider briefly a simpler form of boundary layer equation (Rosenhead, 1963, Ch.V)

$$\left.\begin{aligned} &y''' + yy'' + \beta(1-y'^2) = 0, \qquad \beta > 0, \\ &y(0) = y'(0) = 0, \quad y'(\infty) = 1. \end{aligned}\right\} \tag{6.6}$$

For large x, let $y' = 1 + q$, where q is small. As $x \to \infty$, $y \to x + d$ say. Substituting in the differential equation and neglecting q^2, we obtain

$$q'' + (x+d)q' - 2\beta q = 0. \tag{6.7}$$

The middle term is removed by the substitution $q = z \exp\{-\frac{1}{4}(x+d)^2\}$, giving

$$z'' + \{\frac{1}{2} + 2\beta + \frac{1}{4}(x + d)^2\}z = 0, \tag{6.8}$$

which is essentially the equation for the parabolic cylinder function (Abramowitz and Stegun, Ch. 19). Hence we can obtain the asymptotic form of the solution of (6.6),

$$y \sim 1 + A(x+d)^{-2\beta-1} \exp\{-\tfrac{1}{2}(x+d)^2\}, \tag{6.9}$$

with parameters A and d.

In solving generalized boundary-value problems, parameters occurring in the differential equation give no special difficulty. But in dealing with singular solutions or asymptotic forms it is usually necessary to carry out some detailed analysis of the equation, which may be a complex task for large nonlinear systems (see Murray, 1974, Ch. 6). We will consider analytical results in the next section, and then discuss the practical implementation of generalized matching.

7. SINGULARITIES AND ASYMPTOTIC SOLUTIONS.

The theory of singular points for linear differential equations is fully developed, and the results are available in standard works such as Ince (1926, Ch. 7). For the second-order equation

$$y'' + f(x)y' + g(x)y = 0, \tag{7.1}$$

the necessary and sufficient conditions for a regular singularity at $x = c$ are

$$f(x) = O(x-c)^{-1}, \quad g(x) = O(x-c)^{-2} \tag{7.2}$$

in the neighbourhood of c. When these conditions are fulfilled, a series solution (of Frobenius type) exists near c in the form

$$y = (x-c)^r\{\alpha_0 + \alpha_1(x-c) + \alpha_2(x-c)^2 + \ldots\}, \tag{7.3}$$

where r is not necessarily an integer. The value of r is determined by substituting (7.3) into the differential equation and equating the coefficient of $(x-c)^{r-2}$ to zero. This method always gives at least one solution at a regular singularity; if it does not give two independent solutions, a second one may be obtained by introducing a logarithmic term (Ince, 1926, §16.3). An example is provided by the solutions of the Bessel equation of integer order, near $x = 0$.

The method extends in a simple way to linear equations of higher order, and the only terms which occur at a regular singularity are of the form

$$(x-c)^r \{\ln(x-c)\}^v,$$

where v is an integer. The behaviour at infinity may be considered in a similar way by transforming the independent variable. Putting $t = 1/x$, equation (7.1) becomes

$$\frac{d^2y}{dt^2} + \left\{\frac{2}{t} - \frac{f(1/t)}{t^2}\right\}\frac{dy}{dt} + \frac{g(1/t)}{t^4}\, y = 0 \tag{7.4}$$

and the point at infinity becomes $t = 0$. If (7.4) satisfies the conditions for a regular singularity at the origin, we can obtain local solutions in t of the usual form, which become asymptotic solutions in terms of x.

Unfortunately the theory of regular singularities is rather limited in application; many important linear problems have singular points which are not regular, and different types of solution have to be investigated. In the linear case, it is always possible to find the location of the singular points, because they are given by the singularities of the coefficients in (7.1) (or the analogous form of higher order). For the case of nonlinear equations, the problem of determining the existence and nature of singularities in the solution is much more difficult. If we restrict ourselves to possible singularities at the end-points, we can identify them when we are given explicit boundary conditions, because they occur when the right-hand side of (2.1) does not exist or is not single-valued. We then have to try various types of singular function, e.g. fractional powers, logarithmic terms, exponential terms, until we can fit the given conditions.

Any additional knowledge that we can obtain about the analytical behaviour of the solution may be useful in setting up the parameters.

For equations which are not covered by the Frobenius theory, it is particularly necessary to use numerical checks on the interval in which the singular solution is valid, or on the effective position of the point at infinity. For example, if a singular solution at $x = 0$ is assumed to apply over the range $[0,h]$, the calculation should be performed with two or three different values of h, to verify the consistency of the assumption.

8. SOLUTION BY THE SHOOTING METHOD

There are three main types of generalized parameter, a constant occurring in the differential equation, a coefficient or exponent in the local solution at an end-point, and the length of the range. In principle we can obtain variational equations for any of these cases, and carry out the Newton method for solving (6.1) as described in Chapter 16. However, the form of the variational equations is different for each type of parameter. For example, suppose we consider equation (2.1) over the range $[0,p]$, where p is a parameter. By making the transformation $t=x/p$, we can write the system as

$$\frac{d\underline{y}}{dt} = p\,\underline{f}(pt,\underline{y}) \tag{8.1}$$

over the range $[0,1]$. The equations for $\underline{q} = \frac{\partial \underline{y}}{\partial p}$ are

$$\frac{d\underline{q}}{dt} = \underline{f}(pt,\underline{y}) + pt\,\frac{\partial \underline{f}}{\partial x} + p\left[\frac{\partial f_i}{\partial y_j}\right]\underline{q}. \tag{8.2}$$

The boundary conditions are $q_i = 0$ wherever y_i is given. In terms of the original variable x we have

$$\frac{d\underline{q}}{dx} = \frac{1}{p}\,\underline{f}(x,\underline{y}) + \frac{x}{p}\,\frac{\partial \underline{f}}{\partial x} + \left[\frac{\partial f_i}{\partial y_j}\right]\underline{q}, \tag{8.3}$$

which represents the standard variational equations with two additional terms on the right-hand side.

It is quite possible to obtain all the different modifications of the variational equations, but a simpler and more uniform procedure is to calculate the Jacobian for use in Newton's method by finite differences. If we approximate the derivative $\frac{\partial y_i}{\partial p_j}$ by perturbing the parameter p_j and forming the difference quotient, the method becomes the same for all types of parameter. This gives a straightforward algorithm on which to base a general program for solving boundary-value problems. As noted earlier, the convergence is not usually affected adversely by using differences instead of derivatives. In very sensitive problems we may have to employ multiple shooting (Chapter 16) or perturbation methods to obtain convergence, but generally the use of the variational equations would not make such cases any easier.

A further type of constraint which does not fit into the categories discussed above is a normalization condition on the solution. This is comparatively trivial for linear problems, where the normalization is merely a scaling of the variables. For example, to solve

$$\left.\begin{aligned} &y'' + \lambda f(x)y = 0, \\ \text{with} \quad &y(a) = 0, \quad y(b) = 0, \end{aligned}\right\} \tag{8.4}$$

it is simple to fix $y'(a) = 1$ say, and to take the parameters as $y'(b)$ and λ. The results can be re-scaled after the problem is solved.

A more general type of normalization may lead to greater difficulty. Suppose we want to solve

$$\left.\begin{aligned} &\underline{y}' = f(x,\underline{y}), \\ \text{subject to} \quad &\int_a^b g(\underline{y})\,dx = A, \end{aligned}\right\} \tag{8.5}$$

where A is known. Assuming that the parameters are boundary conditions, we would generally have s-1 given conditions and s+1 unknown parameters to be determined. Shooting and matching gives s nonlinear equations of form (6.1), and the additional equation is the second of (8.5). One possible method of proceeding is to use the matching equations to eliminate s of the parameters, then to write the integral condition in terms of the remaining parameter p_k say. To obtain a correction to the current value of p_k, we can expand this condition to first order as follows

$$\int_a^b \left\{ g(\underline{y}\ (p_k)) + \sum_i \frac{\partial g}{\partial y_i}\frac{\partial y_i}{\partial p_k}\ \delta p_k \right\} dx = A. \tag{8.6}$$

The elimination of the first s parameters need not be done explicitly of course; the boundary-value problem is solved for a certain fixed value of p_k to give $\underline{y}(p_k)$, then re-solved with a perturbed value in order to estimate $\frac{\partial y_i}{\partial p_k}$. These quantities are then used in (8.6) to calculate δp_k. The complexity of the process is obvious, but it generally happens that the integral in (8.5) is fairly stable with respect to variations in y, and so the problem is not too difficult to handle numerically.

19

EXPANSION METHODS

1. INTRODUCTION

Nearly all of the methods described in this book have been finite difference step-by-step methods. We discuss here alternative approaches coming from the field of approximation theory, in which the unknown solution $f(x)$ is expanded in terms of a set of known functions $h_i(x)$:

$$f(x) \simeq f_N(x) = \sum_{i=1}^{N} a_i\, h_i(x). \tag{1.1}$$

The unknowns then being the expansion coefficients a_i (which depend on N, as may the $h_i(x)$). An algorithm based on (1.1) is an expansion method; examples include Galerkin, collocation, least squares and Tchebyshev methods, and these correspond to different criteria for determining the a_i. Methods of this type have advantages in some circumstances over step-by-step methods and disadvantages in others; an example of their use in solving parabolic equations was given in Chapter 14.

2. LINEAR PROBLEMS

Any set of coupled linear ordinary differential equations may be written in more than one way in the form

$$\mathcal{L}\, f(x) = g(x)\ , \tag{2.1}$$

where $f(x)$, $g(x)$ are vector valued functions of x with, say, n components and $\mathcal{L}$ is a differential operator.

For example, the equation

$$\text{(a)} \quad \frac{d^2}{dx^2} f(x) = g(x), \qquad x \in [a,b] , \tag{2.2a}$$

is equivalent to

$$\text{(b)} \quad \left.\begin{aligned} \frac{d}{dx} f_1(x) - f_2(x) &= 0 \\ \frac{d}{dx} f_2(x) &= g(x) \end{aligned}\right\} \text{i.e.} \begin{pmatrix} \frac{d}{dx} & -1 \\ 0 & \frac{d}{dx} \end{pmatrix} \begin{pmatrix} f_1 \\ f_2 \end{pmatrix} = \begin{pmatrix} 0 \\ g \end{pmatrix}, \tag{2.2b}$$

as is obvious on setting $f = f_1$, $\frac{df}{dx} = f_2$.

We shall call the <u>total order</u>, p, of the system the sum of the orders of the components; $p = 2$ in this example. Associated with the differential system will be a set of q boundary conditions; we write these in the form

$$B_i\, f(x) = b_i, \; i = 1,\ldots,q, \tag{2.3}$$

where the B_i are linear (possibly differential) operators defined on a point set $P \in [a,b]$. In most cases we expect $q = p$. We recognise (2.3) as representing a general set of multi-point boundary values; for example, we can append to (2.2) either

(a) $f(a) = 1$, $f'(a) = 1$, giving an initial value problem

or (b) $f(a) + f'(\frac{a+b}{2}) = 0$, $f'(b) = 1$, giving a (rather artificial) boundary value problem.

We consider here expansion methods for the class of problems defined by (2.1), (2.3); the methods as described assume the existence of an unique solution, and attempt to approximate it.

If the function $f(x)$ is vector valued so are the individual terms $a_i\, h_i(x)$. We shall then write a_i and h_i as column vectors, and the 2×2 example above appears as

$$f_N = \begin{bmatrix} f_{1N} \\ f_{2N} \end{bmatrix} = \sum_{i=1}^{N} \begin{bmatrix} h_{1i}(x) \\ h_{2i}(x) \end{bmatrix} \Lambda \begin{bmatrix} a_{1i} \\ a_{2i} \end{bmatrix}$$

and we shall rewrite (1.1) in the form

$$f \simeq f_N = \sum_{i=1}^{N} h_i \,{}_\Lambda\, a_i , \tag{2.4}$$

where Λ denotes element-by-element multiplication:-

$$(p \,{}_\Lambda\, q)_j = p_j \, q_j . \tag{2.5}$$

3. COLLOCATION METHODS

Conceptually the simplest algorithm for determining the n-vectors a_i in (2.4) is that of collocation. If we substitute f_N for f in the boundary conditions (2.3) we obtain the q equations

$$\sum_{j=1}^{N} B_i(h_{j\Lambda} \, a_j) = b_i , \qquad i = 1,\ldots, q \tag{3.1}$$

which represents q×n conditions to be satisfied by the n×N components of $a_1,\ldots,a_N$. If we now choose N-q points $x_k \in [a,b]$, k=1,...,N-q, we can impose the further, "collocation" conditions that the differential equation (2.1) be satisfied at x_k:

$$\mathcal{L} f_N(x_k) = \mathcal{L} \sum_{j=1}^{N} (h_j(x_k) \,{}_\Lambda\, a_j) = g(x_k), \quad k=1,2,\ldots,N-q \tag{3.2}$$

If we are not unlucky, equations (3.1) and (3.2) have an unique solution; and we may hope that as $N \to \infty$, $f_N(x) \to f(x)$ for all $x \in [a,b]$.

4. RESIDUAL MINIMIZING TECHNIQUES

The collocation technique is extremely easy to set up and to use; it satisfies the boundary conditions exactly, and the differential equation exactly on the point set $\{x_k\}$. Its major disadvantage is that the size of this point set is related to N in an uncomfortable way; we take only sufficient points to yield defining equations, and on these points we make the <u>residual</u> <u>vector</u> $r_N(x) = f_N(x) - g(x)$ vanish. There is an obvious analogy with Lagrange interpolation; the possible nonconvergence of a sequence of Lagrange interpolants is well known, and suggests that the behaviour of collocation methods may depend crucially on the choice of point set $\{x_k\}$. In practice this is found to be true. In some circumstances, a choice guaranteed to yield convergence can be found (see chapter 14). In other cases, we can overcome this dependence by minimizing some measure of the overall size of the residual. The most common measures used are:

(a) Least squares:

$$\text{minimize } ||r_N||^2 = \sum_{j=1}^{n} \int_a^b w_j(x)\ (r_N(x))_j\ dx\ , \tag{4.1}$$

where $w_j(x)$ is a positive but otherwise arbitrary weight function, and $(r_N(x))_j$ is the jth component of $r_N(x)$

(b) Minimax:

$$\text{minimi e } ||r_N||_\infty = \max_{1 \leq j \leq n}\ \sup_{x \in [a,b]} |(r_N(x))_j\ w_j(x)|. \tag{4.2}$$

In both cases, it is common to set $w_j(x) = 1$.

(c) In addition, the method of moments, or Galerkin method, is popular. This chooses a second set of functions $\phi_i(x)$, and choose the a_i to satisfy the N-q "moment equations"

$$\int_a^b \phi_i(x)\ {}_\Lambda\ r_N(x)\ w(x)\ dx = 0 \quad i = 1,2,\ldots,N\text{-}q. \tag{4.3}$$

Methods (a) and (c) lead to sets of linear equations for our linear problem, and hence are easier to use than method (b), which yields nonlinear equations. For example, the Galerkin equations (4.3) written out in full have the form

$$\sum_{j=1}^{N}\left[\int_a^b \phi_i(x)_\Lambda \ \ell\, h_j(x)\, w(x)\, dx\right] a_j = \int_a^b \phi_i(x)_\Lambda g(x)\, w(x)\, dx, \quad (4.4)$$

$$i = 1,\ldots,N-q\ ,$$
$$j = 1,\ldots,N\ ,$$

which together with the boundary conditions (3.1) determine in general an unique set of vectors a_j, $j = 1,\ldots,N$. In practice, there is little difference in accuracy between these three methods for a given choice of $\{h_i\}$ and N; there seems little point therefore in spending the extra effort needed to implement (b) unless there are special reasons for requiring a minimax residual (note that a minimax residual does not guarantee a minimax error in the solution).

5. DISCRETIZATIONS; UNIFORM TREATMENT OF THE BOUNDARY CONDITIONS

The methods outlined above have two computational disadvantages. The first is that the boundary conditions are treated exactly, while the differential equation is only approximated. The resulting nonuniformity is only a minor nuisance for ODE's, but requires fixing when the method is extended to more than one dimension, when it is not usually even possible to satisfy the boundary conditions exactly. The second concerns the evaluation of the integrals involved in (4.3), and similarly for the least squares methods. In practice these are replaced by some discretization:

$$\int_a^b w(x)\, f(x)\, dx \rightarrow \sum_{l=1}^{Q} w_l\, f(x_l)$$

and two distinct types of discretization have been used:

(a) A set of Q equally spaced points on [a,b], with $w_l = 1/Q$. (See, e.g., Barrodale & Young (1970)).

(b) The w_l, x_l chosen to represent a Q point quadrature formula. In the second case, there is a great temptation to treat such integrals independently, since it turns out that the difficulty of each increases with i, j; and some programmes known to the author hand over each integral to an automatic quadrature routine. This is hopelessly inefficient in practice, and these algorithms only become competitive when a fixed set of points and weights is used for all of the integrals involved. In this case, choices (a) and (b) lead to formally identical computational procedures, and their relative efficiency is tested, by example, below.

Now if we insert this discretization into the defining equations for the least squares method, we find that we are computing the a_i from the condition that the (weighted) means square residual for the differential equations be minimised over the point set $\bar{Q} = \{x_l,\ l=1,\ldots,Q\}$ subject to the condition that the boundary equations (3.1) be satisfied exactly. We can treat the differential equations and the boundary conditions uniformly by appending the boundary points to $\bar{Q}$ and seeking a (weighted) least squares solution of the whole overdetermined system. That is, we write down the Q + q (vector) equations in N (vector) unknowns

$$\begin{aligned} \mathcal{L} f_N(x_l) &= g(x_l), \quad l = 1,\ldots,Q\ , \\ B_i f_N &= b_i, \quad i = 1,\ldots,q \end{aligned} \tag{5.1}$$

and seek a discrete, weighted least squares solution of these. This solution satisfies the block normal equations

$$A^T WA\, a = A^T d, \tag{5.2}$$

where W is a diagonal matrix of weights, and for $1 \leqslant i \leqslant Q+q$,

$$d_i = g(x_i),\ i \leqslant Q \text{ and } d_i = b_{i-q},\ Q < i,$$

and $\quad A_{ij} = L_\Lambda h_j(x_i), \quad 1 \leqslant j \leqslant N,$

with $\quad L = \mathcal{L}\ ,\ i \leqslant Q$ and $L = B_{i-Q},\ Q < i.$

Similarly, the discretised moment (Galerkin) equations become, after incorporating the boundary conditions

$$\Phi^T A a = \Phi^T d, \tag{5.3}$$

where A,d are as in (5.1) and Φ is the (Q+q) × N matrix

$$\phi_{ij} = \phi_j(x_i), \qquad j = 1,\ldots,N, \quad i = 1,\ldots,(Q+q). \tag{5.4}$$

6. AN EXAMPLE

As an example of the technique, we consider the following problem, taken from Barrodale and Young (1970).

$$f = f^{(iv)}(x) - 3601 f^{(ii)}(x) + 3600 f(x) = 1800x^2-1, \quad 0 \leq x \leq 1$$

$$f(0) = 1;\ f(1) = 1.5+\mathrm{Sinh}(1);\ f'(0) = 1;\ f'(1)=1+\mathrm{Cosh}(1)$$

Exact solution: $f(x) = 1 + 0.5\, x^2+\mathrm{Sinh}(x)$.

Expansion: $$f_N(x)= \sum_{i=1}^{N} a_i\, x^{i-1}.$$

Discretizations: D_a: Q points equally spaced on [0,1]; $w_i = 1/Q$,

D_b: Q Gauss-Legendre points and weights on [0,1].

In each case the boundary conditions are allotted a weight 1. We report two measures of the accuracy:-

1) $$\epsilon_{max} = \max_{x \in \bar{Q}} |f(x) - f_N(x)|,$$

2) $$\epsilon_{LS} = \left\{ \frac{1}{10} \sum_{i=1}^{10} |e_N(y_i)|^2 \right\}^{\frac{1}{2}},$$

where $\{y_i\}$ is a set of 10 equally spaced points in [0,1].

The overdetermined system to be solved has the form

$$\sqrt{w_i}\sum_{j=1}^{N} a_j \quad h_j(x_i) = \sqrt{w_i}\,(1800\ x_i^{\ 2} - 1),\ i = 1,2,\ldots,Q$$

$$a_1 = 1\ ,$$

$$\sum_{j=1}^{N} a_j = 1.5 + \mathrm{Sinh}(1)\ ,$$

$$a_2 = 1\ ,$$

$$\sum_{j=2}^{N} (j-1)a_j \quad = 1 + \mathrm{Cosh}(1).$$

TABLE 1

Choice of Points	N \ Q	2		4		6		8	
		ϵ_{LS}	ϵ_{max}	ϵ_{LS}	ϵ_{max}	ϵ_{LS}	ϵ_{max}	ϵ_{LS}	ϵ_{max}
D_a	5	1.7	2.3	0.72	1.10	-	-	-	-
	10	1.7	2.3	0.73	1.15	0.030	0.032	9.2,-9	1.2,-8
	20	1.7	2.3	0.74	1.16	0.053	0.057	1.6,-7	1.8,-7
	50	1.7	2.3	0.74	1.16	0.068	0.073	2.9,-7	3.2,-7
D_b	4	1.6	2.3	0.74	1.16	-	-	-	-
	8	1.7	2.3	0.74	1.16	0.080	0.086	4.0,-7	4.3,-7
	12	1.7	2.3	0.74	1.16	0.080	0.086	4.0,-7	4.3,-7
max residual, $\lvert r_N\rvert_{max}$	12	3.00		-0.8		-1.1		-3.0, -3	

The errors in the solution are displayed in table 1.

This example illustrates several points. First, in this kind of problem for which the solution is analytic, convergence is extremely rapid in N. Second, as Q increases for fixed N, convergence of ϵ_{LS} and ϵ_{max} is observed; moreover, this convergence is more rapid for the "Integration rule" D_b than for the equally spaced points D_a. Third, there is little difference between ϵ_{max} and ϵ_{LS}, indicating that it is unlikely to be worthwhile minimising $||r_N||_\infty$.

This problem was solved in the Tchebychev norm in Barrodale and Young (1970), using a seventh degree spline with nine knots (that is, seventeen parameters). The result there had $\epsilon_{max} \sim 10^{-8}$, suggesting that in this example at least piecewise polynomials show no advantage over a single polynomial.

7. RELATIONS BETWEEN THE METHODS

The Galerkin, least squares, and collocation methods are clearly closely related. If for Galerkin we choose

$$\phi_i(x) = Lh_i(x)$$

then $\Phi = A$ in (5.3) and the equivalence is complete. Further, if we choose the minimum possible number of points: $Q + q = N$, then A, Φ are square matrices; if they are non-singular both least squares and Galerkin reduce to the solution of the system

$$A a = d,$$

and these are then collocation equations for the same choice of points. These identities hold also if we satisfy the boundary conditions exactly.

8. ERROR ESTIMATES; CHOICE OF THE FUNCTIONS h_i

The error estimates in the table above were derived by knowing the right answer; we are not always so lucky. In practice, we have available only two readily computed parameters which might provide estimates:-

(1) The residual vector $r_N(x)$; we can always stop the process when in some norm $||r_N||$ seems sufficiently small. This procedure has the advantage of being available for any choice of the expansion function $h_i(x)$. It has the wellknown disadvantage that if the problem is at all ill-conditioned, the residuals can be misleading. In the context of linear equations, we are used to being warned that a small residual need not imply a small error. In the current context, because $r_N = \mathcal{L}\, e_N$ and the differential operator $\mathcal{L}$ is unbounded it is often found numerically that $||r_N|| \gg ||e_N||$, and the last line of the table above shows that this happens in the example given; using $||r_N||$ as an error indicator would have grossly underestimated the accuracy of the solution.

(2) The expansion coefficients a_i, $i = 1,\ldots,N$. For a suitable choice of the h_i, these lead as is well known to cheap and (usually) realistic estimates of the error in an L_2 sense. In particular, with

$$e_N = f - f_n \text{ and } f = \sum_{i=1} b_i\, h_i \,,$$

we can define

$$||e_N||_2^2 = \sum_{k=1}^{n} \int_a^b (e_N(x))_k^2 w(x)dx = \int_a^b e_N^T(x)e_N(x)w(x)dx$$

$$= \sum_{\substack{i=1\\j=1}}^{\infty} \sum_{k=1}^{n} (b'_i)_k (b'_j)_k \int_a^b (h_i)_k(x)(h_j)_k(x)w(x)dx, \qquad (8.1)$$

where $b'_i = b_i - a_i, \qquad i \leqslant N,$

$= b_i\,, \qquad i > N.$

Now if the vector functions h_i are orthonomal:

$$\int_a^b h_i(x)h_j(x)w(x)d = \delta_{ij}\, \underline{1}\,,$$

(8.1) takes the simple form

$$||e_N||_2^2 = \sum_{i=1}^{N} (b_i-a_i)^T(b_i-a_i) + \sum_{i=N+1}^{\infty} b_i^T b_i = S_1+S_2. \qquad (8.2)$$

Now if we assume that the a_i are close approximations to the b_i, $i = 1,\ldots,N$, and ignore S_1; and if we further assume that the b_i converge at some power rate:

$$||b_i||_2 \sim Bi^{-r},$$

then we obtain the estimate

$$||e_N||_2^2 \sim \frac{B^2}{2r-1} N^{-2r+1} \sim \frac{N}{2r-1} a_N^2. \qquad (8.3)$$

In (8.3) the only unknown is r; this could be estimated from the observed rate of decrease of the a_i, but the precision of the estimate is hardly such as to warrant this, and we will normally set $2r-1 = 1$.

More detailed estimates (see, e.g., Freeman and Delves (1974), Musa and Delves (to appear) show that indeed S_1 is at most $O(S_2)$), so that an estimate of the form (8.3) is applicable with perhaps a larger numerical coefficient; and that in favourable circumstances we should find

$$||b_i||_2 \sim B\,C^i,$$

that is, an exponential convergence rate; such a rate is in fact exhibited by the numerical example given above.

These possibly very high rates of convergence are the major advantage of the use of a polynomial basis for the solution. If the solution of the problem has low analytic continuity (discontinuous derivative of some low order on [a,b]) the parameter r in (8.3) will be small and convergence slow; further, for finite N, it is impossible to distinguish a discontinuous derivative from a very rapidly changing one, so that low apparent convergence rates will obtain for problems leading to such solutions also. It is therefore often suggested that piecewise polynomials will yield better results.

For partial differential equations, this suggestion (leading to the finite element method) has been very successful. For ODE's, the evidence to date is much less clear cut. This is probably because expansion methods are less often used, so that the experience is not yet there; and partly because a major advantage of the finite element method in more than one dimension is its ability to handle awkward shaped regions efficiently. In one dimension, regions do not have awkward shapes. My own feeling is that we will normally get the highest efficiency by using a high degree polynomial, but that if the going gets too tough one should subdivide the region $[a,b]$ into two, or a few, such intervals, thus effectively using a high degree piecewise polynomial but only when it seems advantageous. A routine of this type exists in the NAG library (but is not, it is fair to say, particularly popular).

9. EXTENSIONS AND CONCLUSIONS

1) Nonlinear equations

Although the formalism above dealt only with linear equations, nonlinear equations can be handled in the same way, provided that the resulting nonlinear algebraic equations can be solved iteratively. The iterative solution in general requires the repeated calculation of a matrix of integrals, and hence it is often felt that an expansion method must be expensive in such cases, but this is not borne out in practice except for the special case of initial value problems for which so many alternative techniques exist; nonlinear boundary value problems are surprisingly expensive to solve by any method.

2) Infinite Intervals

Expansion methods also have the advantage over both shooting and implicit finite difference methods that problems defined on infinite intervals ($[0,\infty)$, $(-\infty, \infty)$), cause no particular difficulties. On $[0,\infty)$, for example, we can use an expansion set such as the Laguerre functions $L_n(\alpha x)e^{-\alpha x}$, or a set such as $1/(x+\alpha)^n$; the formalism then applies unchanged.

More details of these methods are given in the books by Fox and Parker (1968), Mikhlin and Smolitsky (1967) and Mikhlin (1971).

We can summarise the main properties of expansion methods, then, as follows:

<u>For</u>

1) They yield a compact solution form which can be transported readily for use elsewhere. This is sometimes an overriding consideration; for example, in producing library approximations for special functions.

2) They handle arbitrary multi-point boundary conditions and other "awkward" types of problems with no difficulty.

3) In favourable circumstances, they converge extremely rapidly; this makes these methods especially attractive when high accuracy is needed in the solution of a boundary value problem.

<u>Against</u>

4) Compared with finite difference techniques, little working experience with these methods exists; and hence, few standard programmes have been written.

5) It is significantly harder to write an efficient standard programme than for a finite difference method. Hence users who write their own are sometimes disenchanted with the results.

<u>Headmasters Report</u>: This method tries hard, and shows flashes of brilliance. May do well in future.

PART 4

FUNCTIONAL DIFFERENTIAL PROBLEMS

20

INITIAL VALUE PROBLEMS FOR DELAY-DIFFERENTIAL EQUATIONS

1. INTRODUCTION

We will be concerned here with some numerical methods for the approximate solution of equations of the form

$$y'(x) = f(x, y(x), y(x-\tau(x))), \quad (x > x_0) \tag{1.1}$$

given certain initial conditions, where $\tau(x) \geqslant 0$. Such problems arise in a wide variety of practical circumstances ranging from infection growth and models of the heart-lung complex, to the design of an overhead collection system for an electric locomotive and machine tooling, whilst applications in analytic number theory and control theory also arise.

Let us observe that equation (1.1) is frequently discussed in the context of the more general equation

$$y'(x) = f(x, y(x), y(x-\tau_1(x)), y(x-\tau_2(x)), \ldots, y(x-\tau_q(x))) \tag{1.1a}$$

where $\tau_i(x) \geqslant 0$ for $i = 1, 2, \ldots, q$.

Whilst there is undoubted interest in the solution of delay-differential equations, the development of numerical techniques does not appear to have progressed to the stage reached with algorithms for the initial-value problem for the equation $y' = f(x, y)$. At the time of writing (though the work of Neves (1975) is relevant) there is not a wide choice of competitive all-purpose coded algorithms for delay equations.

A theory for numerical methods for delay-differential equations does nevertheless exist, and a treatment of consistency, convergence, and stability, and stability regions, may be found in the growing literature. We have endeavoured to be fairly comprehensive in the references which we have collected for inclusion here. Since the general bibliography covers a wide field, we mention in section 5 the contributors to the topic presently under discussion.

It is appropriate to draw attention, here, to some existing surveys of numerical methods for delay-differential equations which have already appeared. In particular we mention El'sgol'ts and Norkin (1973), who include a brief survey, and Cryer (1972), Tavernini (1971). The thesis of Wiederholt (1970) opens with a survey. The paper of Fox, Mayers, Ockenden and Tayler (1971), though concerned with a specific problem, is also valuable in this context, since Part B of this paper covers some suggested techniques with wider application.

Before proceeding, we should elaborate on the nature of the problem and indicate the limitations of our discussion. We shall consider the first order equation (1.1) with the restriction $\tau(x) \geqslant 0$; such equations (particularly, with $\tau(x) > 0$) are termed delay or retarded ordinary differential equations¶. In the case $\tau(x) < 0$, the equation is called an advanced ordinary differential equation, whilst other types of equation involving a deviating argument, such as $y'(x) = F(x, y(x - \tau(x)), y(x), y(x + \tau(x))$, are usually called neutral equations. (It is not clear to us that the terminology is completely standard.) We shall be concerned with delay equations. Furthermore, we shall restrict attention to initial-condition problems for first order equations, and direct methods suitable for use on a computer. (A shooting method for a boundary-value problem for a second order equation was used by de Nevers and Schmitt, 1971.) We shall not concern ourselves with semi-analytical techniques in which an approximate method is employed to evaluate an explicit form of the solution.

2. BASIC NOTIONS

There now exists a number of texts outlining the basic theory of difference-differential equations, of which (1.1) forms an example. In particular we mention El'sgol'ts and Norkin (1973) (being a translation from the 1970 Russian edition), which contains a substantial bibliography, Myskis (1949), and Bellman and Cooke (1963); we refer also to Schmitt (1972).

Whereas the required solution of the equation $y' = f(x, y)$ $(x>x_0)$ is defined uniquely by stating the value $y(x_0)$, the initial conditions determining the solution of (1.1) generally involve prescribed values

¶We shall refer to $\tau(x)$ as the delay, or lag. Others sometimes apply these terms to $x-\tau(x)$.

of $y(x)$ on an interval. The precise nature of these initial conditions depends upon the nature of the lag, and the range of values of the argument. If the solution of the problem is defined by an initial value at a single point then we shall refer to the problem, with Feldstein (1964), as an initial value retarded ordinary differential equation. In general, if we seek the solution of (1.1) for $x_0 \leqslant x \leqslant X$, then we require initial information of the form

$$y(x) = y_0(x) \quad (\min(x_0, x_*) \leqslant x \leqslant x_0),$$

where $x_* = \min_{x\varepsilon(x_0,X]} \{x - \tau(x)\}$. The following examples serve to illustrate.

Example 2.1. An equation with constant lag, of the form $y' = f(x, y(x), y(x-\alpha))$, $\tau(x) \equiv \alpha > 0$ is given by $y'(x) = \lambda y(x-\alpha)$, $x > x_0$ with prescribed conditions $y(x) = \gamma$ for $x_0-\alpha \leqslant x \leqslant x_0$. The solution is

$$y(x) = \gamma \sum_{r=0}^{\left[\frac{x-x_0}{\alpha}\right]+1} \{\lambda^r (x-x_0 - (r-1)\alpha)^r / r!\}$$

where $[z]$ denotes the integer part of z.

Example 2.2. An example of an equation with decreasing delay as the argument increases is (Bellman, Buell, and Kalaba 1965) $y'(x) = -y(x - e^{-x}-1) + \{\cos x + \sin(x - e^{-x}-1)\}$. To determine a solution for $x > 0$ we specify $y(x) = \sin x$ for $x \varepsilon [-2, 0]$; the solution is then $y(x) = \sin x$ for $x \geqslant 0$.

Example 2.3. If $\min\{x-\tau(x)\} = x_0$, then (1.1) presents an initial-value retarded differential equation. The problems of the form

(i) $y'(x) = ay(\lambda x) + by(x) \quad (x \geqslant 0;\ 0 < \lambda < 1)$

$y(0) = y_0$

considered by Fox et al (1971) furnish examples. Further illustration is given by examples of Feldstein (1964), of form

(ii) $y'(x) = y(\tfrac{1}{2}x) + y(x^2)$,

(iii) $y'(x) = y(\tfrac{1}{2}x)$, and (iv) $y'(x) = y(x^2)$.

Observe that (ii) is a "delay equation" only if $|x| \leqslant 1$, and then provides an example of an equation (1.1a) with more than one lag.

Example 2.4. Feldstein and Goodman (1971) give a number of functional differential equations; in particular $y'(x) = y(x-1)-3y(x-\sqrt{2}) + y(x-\pi)$ $(x > 0)$ with $y(x) = 1$ for $x \leqslant 0$ provides an example of a delay equation with more than one delay.

Example 2.5. The equations (Feldstein 1964) $y'(x) = y(x^2)\,(0 \leq x \leq 1)$, $y'(x) = y(\frac{1}{2}(1+x))\,(1 < x)$, with $y(0) = 0$ illustrate a problem which will be raised later. Note that the lag assumes different forms in the two intervals, but is continuous.

Remark: In some cases, the lag $\tau(x)$ in an equation (1.1) with deviating argument depends upon the solution, and it is not possible to determine, in advance, the range of the initial conditions required. El'sgol'ts and Norkin (1973, p.43 and p.54, etc.) give a number of examples of this type where discontinuities in the initial data have "adverse" effects on the solution, causing lack of smoothness, non-uniqueness, or non-existence over intervals, of the solution to the problem.

It will be observed, from Example 2.1, that even in a simple example of a retarded equation the solution may suffer discontinuities in its derivatives. Thus in the special case where $\lambda = \alpha = 1$ (Bellman and Cooke 1963, p.45) the solution of the equation $y'(x) = y(x-1)$ for $x \geq 1$ with $y(x) = 1$ for $0 \leq x \leq 1$ is obtained by a "continuation process" giving $y(x) = x$ for $1 \leq x \leq 2$, $y(x) = x + \frac{1}{2}(x-2)$ for $2 \leq x \leq 3$, $y(x) = x + \frac{1}{2}(x-2)^2 + \frac{1}{3}(x-3)^3$ for $3 \leq x \leq 4$, etc. (The continuation process is referred to in El'sgol'ts and Norkin (1973) as "the method of steps", and it provides some insight, as shown below, into analytical and numerical methods for the solution of certain delay equations.) The point to emphasise at present is the discontinuity of $y'(x)$ at $x = 1$, of $y''(x)$ at $x = 2$, of $y'''(x)$ at $x = 3$, and so on. The presence of such discontinuous derivatives limits the usefulness of numerical formulae which are designed to achieve high accuracy for functions with high-order derivatives. However, in the example cited the solution becomes progressively smoother with increasing argument, and the points of possible discontinuity are known in advance. (It will also be noted that in Example 2.2, the solution is analytic!)

3. INTRODUCTION TO THE USE OF NUMERICAL METHODS

Most of the numerical methods for initial-value problems for ordinary differential equations can be adapted to give corresponding techniques for delay equations. The range of methods therefore comprises one-step methods (including Kuler's method and Runge-Kutta type methods), multistep methods and block-implicit methods. In each of these methods the standard formula must generally be augmented by an interpolation formula (see below).

The deferred approach to the limit may be employed in some circumstances, and in Fox et al (1971) we see such a case; the problem studied there (see Example 2.3) can also be converted to an integral equation

$$y(x) = (a/\lambda)\int_0^{x\lambda} y(t)\,dt + b\int_0^x y(t)\,dt + y_0, \quad (\lambda < 1),$$

which, for a region $0 \leqslant x \leqslant X$, can be regarded as a Fredholm equation with a discontinuous kernel. (Corresponding integral equation techniques, including product integration methods, collocation, and the τ-method, are then applicable.)

Let us recall that the solution of a delay equation may possess discontinuous derivatives, but let us nevertheless investigate how automatic methods for ordinary differential equations might be adapted to deal with delay equations. To begin, we consider the case $y'(x) = y(x-1)$, $x \geqslant 1$ ($y(x) = 1$ for $0 \leqslant x \leqslant 1$), discussed in the final paragraph of §2. We shall contrast and relate the numerical and analytical methods of solution. The analytical method of solution proceeds in steps, in the first of which we obtain the solution in $[1, 2]$. Thus, for $1 \leqslant x \leqslant 2$ the equation and initial condition give $y'(x) = 1$ $(1 \leqslant x \leqslant 2)$ and we have an initial value problem to solve analytically, or numerically with an automatic program, respectively. The exact solution is $y(x) = x$ for $1 \leqslant x \leqslant 2$; a numerical method would generally supply approximate values $\tilde{y}_r = \tilde{y}(z_r)$ of the solution $y(x)$ approximated by $\tilde{y}(x)$ at discrete points $z_r \in \{z_s | 1 \leqslant z_s \leqslant 2\}$. On the interval $2 \leqslant x \leqslant 3$ we have, in the analytical solution, the problem $y'(x) = y(x-1) = x - 1$, with $y(2)$ given. For the approximate numerical solution we have the nearby problem $y'(x) = \tilde{y}(x-1)$, $x \in [2, 3]$, with $y(2)$ given as $\tilde{y}(2)$. Any algorithm for the numerical solution of this problem may need a value $\tilde{y}(x-1)$ not amongst the computed values $\{\tilde{y}(z_r) | z_r \in [1, 2]\}$ and such a value would have to be obtained by interpolation or extrapolation. This stage of computing the solution in $[2, 3]$ is typical and we have illustrated a key feature of the numerical solution of delay equations: the general need to interpolate or extrapolate on computed values of the solution. Moreover, we have seen that the function being interpolated is an approximation to a function $y(x)$ with discontinuous derivatives of some order. The location of these discontinuities should govern the choice of interpolation formula; in polynomial interpolation for $\tilde{y}(x)$ with $2 \leqslant x \leqslant 3$ we should employ $\tilde{y}(z_r)$ with $2 \leqslant z_r \leqslant 3$ for example. Discontinuities might also affect the extrapolation problem; however, the solution becomes progressively smoother with increasing x, and the interpolation problem becomes correspondingly less difficult. We shall elaborate on this approach later, but the need for interpolation will be seen to impose difficulties on any simple implementation of an automatic differential equation routine. However, Neves (1975) seems to have achieved some success in this direction.

The equation with constant lag is a special one, in particular since the location of any discontinuities can be determined in advance. For equation (1.1) there will in general be jumps in derivatives at points x_n such that

$$x_n - \tau(x_n) = x_s, \qquad (n \geqslant 1,\ s \leqslant n - 1)$$

with x_0 given by the initial condition for (1.1). Additional points of discontinuity may, however, exist. Observe that if $\tau(x)$ is monotonic then

$$x_n - \tau(x_n) = x_{n-1} \quad (n \geqslant 1). \tag{3.1}$$

It is convenient to consider only this case in the following discussion.

The process of continuation may fail even if executed analytically. Such cases of failure are termed "singular cases" (El'sgol'ts and Norkin 1973, p.11), and they cannot occur if $\inf_x \tau(x) \geqslant \tau_* > 0$; the difficulty occurs if $\lim_{n\to\infty} x_n = \overline{x}$ in (3.1), since the solution in steps cannot then proceed past $x = \overline{x}$. Example 2.5 furnishes such a case, with $\overline{x} = 1$. The continuation process outlined for the numerical or analytical solution can however be extended (though complications may arise) to cover "non-singular" equations with varying lag (Cryer 1972, p.56) and in the numerical method we may have the opportunity to use automatic procedures for initial value problems for ordinary differential equations. We again refer to the work of Neves.

Another case where we can exploit automatic procedures is given when the lag $\tau(x)$ in (1.1) is monotonic decreasing, though not necessarily strictly decreasing (Bellman, Buell and Kalaba 1965) with $\tau(x) > 0$. (See Example 2.2.) The technique of Bellman <u>et al</u> eliminates the need for interpolation (as required above). Defining the sequence $\{x_n\}$ by means of (3.1), the solution on $x_r \leqslant x \leqslant x_{r+1}$ is given by the continuation process in terms of previous values on $[x_{r-1}, x_r]$. We suppose that the solution has been found for $x_0 \leqslant x \leqslant x_s$, and we define $\gamma_0(x) = x$, $\gamma_1(x) = x - \tau(x)$, $\gamma_2(x) = \gamma_1(\gamma_1(x)), \ldots, \gamma_s(x) = \gamma_1(\gamma_{s-1}(x))$. We then formulate, on $[x_s, x_{s+1}]$, a system of differential equations for the functions

$$\left.\begin{aligned} &y_s(x) = y(x),\ y_{s-1}(x) = y(\gamma_1(x)),\\ &y_{s-2}(x) = y(\gamma_2(x)),\ldots,\ y_0(x) = y(\gamma_s(x)).\end{aligned}\right. \quad x_s \leqslant x \leqslant x_{s+1} \tag{3.2}$$

As x varies in $[x_s, x_{s+1}]$, $\gamma_r(x)$ varies through $[x_{s-r}, x_{s-r+1}]$, and the original equation (1.1) now yields

$$\frac{d}{d\gamma_{s-k}}\{y_k(\gamma_{s-k})\}=f(\gamma_{s-k}(x),y_k(x),y_{k-1}(x)) \text{ with } y_k(x_k)=y(x_k), x_s\leqslant x\leqslant x_{s+1},$$

for $k = 0, 1, \ldots, s$. From these equations we obtain, for $x\varepsilon[x_s, x_{s+1}]$,

$$\left.\begin{aligned} &y_s'(x) = f(x, y_s(x), y_{s-1}(x)), \quad y_s(x_s) = y(x_s),\\ &y_{s-1}'(x) = \gamma_1'(x)f(\gamma_1(x), y_{s-1}(x), y_{s-2}(x)), y_{s-1}(x_s)=y(x_{s-1})\\ &\qquad\cdots\\ &y_0'(x)=\gamma_s'(x)f(\gamma_s(x),y_0(x),y(\gamma_{s+1}(x))),y_0(x_s)=y(x_0)\end{aligned}\right\} \tag{3.3}$$

In (3.3), $y(\gamma_{s+1}(x))$ is a given initial function defined on $[x_{-1}, x_0]$. Integration of the system (3.3) may be achieved with an automatic programme for a system of first order equations. The solution provides $y_s(x) = y(x)$ for $x_s \leqslant x \leqslant x_{s+1}$, and we can proceed to the interval $[x_{s+1}, x_{s+2}]$, the number of equations in the system (3.3) rising by one.

4. STEP-BY-STEP FORMULAE FOR EQUATIONS WITH DELAY

In § 3 we indicated two approaches whereby procedures for equations without delay might possibly be adapted for equation (1.1) with delay. We now show how basic formulae, which form the foundations of methods for the initial-value problem $y' = f(x, y)$, can be adapted or modified for the treatment of a delay term. This gives insight into the interpolation problem raised in §3, though the discussion here is at the basic level with no comment on the possibility of the adaptive use of formulae.

A primitive formula which has been studied for (1.1) is that of Euler's method. If $y'(x)$ is bounded and Riemann-integrable, $y(x+h) - y(x) = h\, y'(x) + o(1)$; the $o(1)$ term is $O(h^2)$ if $y'(x)$ is Lipschitz continuous. From this relation we construct Euler's formula for equation (1.1), namely

$$\tilde{y}(x + h) - \tilde{y}(x) = h \;\; f(x, \tilde{y}(x), \tilde{y}(x - \tau(x))), \text{ for } x \geq x_0. \qquad (4.1)$$

Choosing† a sequence $x_{r+1} = x_r + h_r$ $(r \geq 0)$, and substituting $h = h_r$, $x = x_r$ in (4.1), we wish to compute values $\tilde{y}_r = \tilde{y}(x_r)$ satisfying

$$\tilde{y}_{r+1} = \tilde{y}_r + h_r f(x_r, \tilde{y}_r, \tilde{y}(x_r - \tau_r)) \qquad (4.2)$$

where $\tau_r = \tau(x_r)$. If $x_r - \tau_r \notin \{x_s\}_0^r$, then we require $\tilde{y}(x_r - \tau_r)$, and the use of piecewise - constant interpolation suggests the substitution

$$\tilde{y}(x_r - \tau_r) := \tilde{y}(x_q) \quad \text{where} \quad x_r - \tau_r \in (x_q, x_{q+1}), \qquad (4.3a)$$

whilst piecewise-linear interpolation would yield

$$\tilde{y}(x_r - \tau_r) := \{\sigma_r''\tilde{y}(x_q) + \sigma_r'\tilde{y}(x_{q+1})\}/h_q, \qquad (4.3b)$$

where

$$\sigma_r' = (x_r - x_q) - \tau_r, \quad \sigma_r'' = h_q - \sigma_r'. \qquad (4.4)$$

Feldstein (1964) (whose study is restricted to initial-value retarded equations), calls the use of (4.3a) with (4.2) the "simplified Euler method"; his more elaborate "customary Euler algorithm" employs the approximations

$$\left.\begin{aligned} \tilde{z}_r \equiv \tilde{y}(x_r - \tau_r) &:= \tilde{y}_q + \sigma_r' f_q \\ f_q &:= f(x_q, \tilde{y}_q, \tilde{z}_q), \\ \tilde{y}_{r+1} &:= \tilde{y}_r + h_r f_r, \end{aligned}\right\} \qquad (4.5)$$

with $\tilde{y}_0 = \tilde{z}_0 = y(0)$, and where q is such that $x_r - \tau_r \in (x_q, x_{q+1}]$. This formula involves two applications of (4.1) with $h = h_r$ and $h = \sigma_r'$ respectively. The order of calculations is $\tilde{y}_0, \tilde{z}_0, f_0, \tilde{y}_1, \ldots, f_r, \tilde{y}_{r+1}, \tilde{z}_{r+1}$ and values of x_r, f_r and $\tilde{y}_r$ are kept for later use (with possibly

† The sequence $\{x_r\}$ is in general different from that in (3.1).

large storage requirements). It is interesting to observe that the sequence $\{f_r\}$ need not be stored if we interpolate with (4.3b) in (4.2), and that this process is then equivalent to using (4.5), as may be seen on reflection.

For equations with varying lag, El'sgol'ts has proposed the use of (4.2) and (4.3a) with mesh $\{x_r\}$ defined by (3.1) (at least we could employ a finer subdivision). This generally avoids jumping across a point of discontinuity in the derivative, but this and related strategies, fail on singular equations (see Example 2.5).

Theoretical results have been established (Feldstein 1964) for a number of variants of the Euler algorithm applied with constant step $h_r = h$, to initial-value retarded equations, as $h \to 0$. The work of Tavernini (1971) includes a more general situation. We also note that the Euler algorithms can be used to solve general delay equations, given suitable starting conditions.

The trapezium rule gives a simple formula with a degree of added sophistication. With minimum conditions, $y(x + h) - y(x) = \frac{1}{2}h\{y'(x)+y'(x+h)\} + o(1)$, and the $o(1)$ term is $O(h^2)$ if $y'(x)$ is Lipschitz continuous and $O(h^3)$ if $y''(x)$ is Lipschitz continuous. Inspired by this, and given (1.1), we may set

$$\begin{aligned} y(x+h) - y(x) \simeq \tfrac{1}{2}h\{&f(x, y(x), y(x-\tau(x))) + \\ &f(x+h, y(x+h), y(x+h - \tau(x+h))\}. \end{aligned} \tag{4.6}$$

Given a mesh x_r with $h_r = x_{r+1} - x_r$ we can obtain values $\tilde{y}_r = \tilde{y}(x_r)$ from the relations

$$\tilde{y}_{r+1}-\tilde{y}_r = \tfrac{1}{2}h\{f(x_r,\tilde{y}_r,\tilde{y}(x_r-\tau_r)+f(x_{r+1},\tilde{y}_{r+1},\tilde{y}(x_{r+1}-\tau_{r+1}))\} \tag{4.7}$$

for $r = 0, 1, 2,\ldots$, when the values $\tilde{y}(x_r - \tau_r)$ and $\tilde{y}(x_{r+1} - \tau_{r+1})$ are obtained by interpolation on values $\{y_s\}$. In contrast with (4.2), we need to <u>extrapolate</u> if $\tau_{r+1} < h_r$.

We shall consider the extrapolation problem later. For the interpolation problem Fox <u>et al</u> (1971) employ either (4.3b) or the piecewise-cubic (simulated Hermite) interpolation formula

$$\tilde{y}(x_r - \tau_r) := \alpha_r y_q + \beta_r y_{q+1} + \gamma_r f_q + \delta_r f_{q+1}, \tag{4.8}$$

where $x_r - \tau_r \in (x_q, x_{q+1})$; from this we obtain $y'(x_r) = f_r$ where $f_s = f(x_s, \tilde{y}_s, \tilde{y}(x_s - \tau_s))$ $(s = 1, 2, \ldots)$. The values y_r and f_r are stored for subsequent use. In (4.8), $\alpha_r = (1+2s_r')(s_r'')^2$, $\beta_r = (1-2s_r'')(s_r')^2$, $\gamma_r = h_q s_r'(s_r'')^2$ and $\delta_r = -h_q s_r''(s_r')^2$, where $s_r' = \sigma_r'/h_q$ and $s_r'' = \sigma_r''/h_q$.

If $x_{r+1} - \tau_{r+1} < x_r$ then (4.8) can be used, with r replaced by $r+1$ throughout, to give $\tilde{y}(x_{r+1} - \tau_{r+1})$. Otherwise, $x_{r+1} - \tau_{r+1} \geqslant x_r$ and we may use extrapolation, or (for example) substitute the formula (4.8) with $q = r$ into (4.7) to give the non-linear coupled equations

$$\left.\begin{aligned} y_{r+1} &= y_r + \tfrac{1}{2}h\, f_r + \tfrac{1}{2}h\, f(x_{r+1}, y_{r+1}, \alpha_r y_r + \beta_r y_{r+1} + \gamma_r f_r + \delta_r f_{r+1}) \\ f_{r+1} &= f(x_{r+1}, y_{r+1}, \alpha_r y_r + \beta_r y_{r+1} + \gamma_r f_r + \delta_r f_{r+1}) \end{aligned}\right. \qquad (4.9)$$

which must generally be solved iteratively. A predictor-corrector technique may be suggested*; for the problem considered by Fox et al, however, equations (4.9) can be solved explicitly because of the nature of $f(x, y, z)$.

Correcting the Euler value using the trapezium rule leads, for the equation $y' = f(x, y)$, to an explicit two stage Runge-Kutta formula: $\tilde{y}_{r+1} - \tilde{y}_r = \tfrac{1}{2}(k_0^{[r]} + k_1^{[r]})$ where $k_0^{[r]} = h_r f(x_r, \tilde{y}_r)$ and $k_1^{[r]} = h_r f(x_r + h_r, \tilde{y}_r + k_0^{[r]})$. The use of a Runge-Kutta formula (of higher order) has been proposed for delay equations by Opplestrup (1973) and it is appropriate to ask what form an extension should take.

Our discussion will be based on the above two stage method, for simplicity. One 'natural' extension to deal with (1.1) is provided by the formulae

$$\left.\begin{aligned} x_{r+1} &= x_r + h_r\ , \\ k_0^{[r]} &= h_r f(x_r, \tilde{y}_r, \tilde{y}(x_r - \tau_r)), \\ k_1^{[r]} &= h_r f(x_r + h_r, \tilde{y}_r + k_0^{[r]}, \tilde{y}(x_{r+1} - \tau_{r+1})), \\ \tilde{y}_{r+1} &= \tilde{y}_r + \tfrac{1}{2}\{k_0^{[r]} + k_1^{[r]}\}, \end{aligned}\right\} \qquad (4.10)$$

using an interpolation formula like (4.3b) or (4.8) to obtain $\tilde{y}(x_r - \tau_r)$ and $\tilde{y}(x_{r+1} - \tau_{r+1})$. If $\tau_{r+1} < h_r$, however, we have the

*We could employ (4.1) with $x = x_r$, $h = x_{r+1} - x_r + \tau_r$ to give simpler formula than (4.9), and such a formula could be used as a predictor.

extrapolation problem associated with the trapezium rule (and to express the required value in terms of $\tilde{y}_{r+1}$ would result in an implicit formula). Indeed, it is arguable that our method should reduce to the standard Runge-Kutta method if $f(x, y(x), y(x - \tau(x)))$ has the form $\phi(x, y(x), y(x))$. With this in mind, we might replace the value of $k_1^{[r]}$ in (4.10) by

$$k_1^{*[r]} := \begin{cases} h_r f(x_r+h_r, \tilde{y}_r+k_0^{[r]}, \tilde{y}(x_{r+1}-\tau_{r+1})), & \text{if } \tau_{r+1} > h_r, \\ h_r f(x_r+h_r, \tilde{y}_r+k_0^{[r]}, \tilde{y}(x_r)+s_r' k_0^{[r]}), & \text{if } \tau_{r+1} < h_r, \end{cases}$$

where $s_r' = (x_{r+1} - \tau_{r+1} - x_r)/h_r = \sigma_r'/h_r$. Recall that a Runge-Kutta formula is devised to give a certain order of local truncation error as $h_r \to 0$, and in delay equations we have the possibility that $\tau(x) \to 0$ as $x \to \bar{x}$; values of x can therefore arise where $|\tau(x)| = O(h_r)$; if we return to basic principles we find the possibility of a number of variants in the extension of Runge-Kutta formulae to delay equations, depending upon the extrapolation formulae employed.†

The usual derivation of Runge-Kutta formulae involves matching with a Taylor series method assuming high-order differentiability, and it is interesting to note the necessary modification to the truncated Taylor series (El'sgol'ts and Norton, 1973 p.239) caused by jumps in derivatives at known points*.

Similar modifications to the Adams multistep formulae (<u>ibid</u> p.240) have been proposed by Zverkina and involve determining the size of jump discontinuities in the derivatives; see Hutchinson (1971 p.4). In the case where derivatives are smooth we can readily adapt the Adams formulae to treat delay equations with constant lag (Zverkina 1968).

Work on the analogue of multistep methods for delay equations has been performed by Tavernini in his doctoral thesis (see Cryer 1972, p.78), and by Hutchinson (1971). The work is too substantial for much more than a reference here, but some points will be emphasised.

The extension of multistep methods studied by Tavernini has the form (with $0 < t < 1$)

$$\{\alpha_k \tilde{y}(x_{n+k-1}+th) + \sum_{i=0}^{k-1} \alpha_i(t)\tilde{y}_{n+i} = h \sum_{i=0}^{k} \beta_i(t) f(x_{n+i}, \tilde{y}_{n+i}, \tilde{y}(x_{n+i}-\tau_{n+i}))$$

---

†Note the results of Feldstein and Goodman (1973) in this context.
*See, also, Hutchinson (1971, p.15).

with $x_r = x_{r-1} + h$, $y(x)$ being presented by the initial conditions for $x \leqslant x_0$. Conditions are imposed on the coefficients $\alpha_i(t)$ and $\beta_i(t)$ which ensure that the function $\tilde{y}(x)$ is continuous, and the theory permits the replacement, by an approximation, of the value $f(x_{n+i}, \tilde{y}_{n+i}, \tilde{y}(x_{n+i} - \tau_{n+i}))$ (presumably so that it is sufficient to compute $\tilde{y}_1, \tilde{y}_2, \tilde{y}_3, \ldots$, and interpolate). For such schemes, the analogue of the Dahlquist theory, relating zero-stability and consistency to convergence, can be established. Cryer (1972) reviews much of the work of Tavernini.

Hutchinson (1971), in a thesis which is also of interest for other aspects, treats the extension of the Dahlquist theory to multi-step formulae with uniform step h, adapted to account for jump discontinuities at known abscissae (and thus preserve high-order accuracy). As Hutchinson comments, one of the difficulties with his modified multi-step methods is the complexity of the modification to deal with the jump discontinuities, and he presents a two-point method based on a degenerate spline, which seems fairly attractive in use (Hutchinson, 1971 p.48 ff) provided the problem is not a singular one. (In the two point method, the points of discontinuity lie on the grid on which the solution is computed.)

5. ADDITIONAL CONSIDERATIONS AND COMMENTS

In our description of algorithms we have seen the need to interpolate, and we generally match the interpolation formulae to the integration method. Spline, Hermite, and Lagrangian formulae are mentioned by Wiederholt (1970), who suggests that the interpolation formula should have the same order as the integration scheme, though (with a fixed step scheme) an interpolation method of order one less is admissible, but not recommended. Kemper (1975) employs a degenerate spline method.

Fox et al (1971) show the effect on their problem of two alternative forms of interpolation, and note the lack of smoothness which they find in the approximate solution using low-order interpolation. (This adversely affects the possibility of h^2-extrapolation.) The remarks of Wiederholt (1970) and of Feldstein and Goodman (1973) appear relevant to this observation. The latter note that the lag term combines with round off error to cause jumps in the computed solution.

We also note, in the context of error propagation, that the stability regions of algorithms for (1.1) differ from those where the methods are applied to an equation with no lag. Wiederholt (1970) makes a study of absolute and relative stability regions, and concludes that there is a significant change in the stability regions when a lag is introduced. Hutchinson (1971) mentions, briefly, the problem of A-stability. The interested reader is directed to these theses, and also to more recent work of Cryer (1973, 1974) on 'highly stable multistep methods'.

We have been unable to give, here, more than the flavour of an intriguing topic. The interested reader may consult the relevant work (listed in the references) by Ball, Barwell, Bellman and his co-workers, Brayton and Willoughby, Chartres and Stepleman, Chosky, Cooke and List, Cryer, de Nevers and Schmitt, El'sgol'ts, El'sgol'ts and Norkin, Feldstein, Fox and his co-workers, Goodman, Gorbunov and Budak, Hill, Hutchinson, Kemper, Neves, Oppelstrup, Schmitt, Tavernini, Thompson, Van de Lune and Wattel, Wiederholt, and Zverkina, in particular.

21

INITIAL VALUE PROBLEMS FOR VOLTERRA INTEGRO-DIFFERENTIAL EQUATIONS

1. INTRODUCTION

The subject of our study here is the numerical treatment of equations of the form

$$y'(x) = f(x, y(x), \int_0^x K(x,t, y(t))dt), \quad (0 \leq x \leq X) \tag{1.1a}$$

with prescribed initial condition

$$y(0) = y_0. \tag{1.1b}$$

Surveys which include some reference to this problem have been given by Cryer (1972) and Baker (1974).

Equation (1.1a) is sometimes referred to as an infinite-delay equation, since (in general) $y'(x)$ depends upon all past values of $y(x)$. Like the delay equations of the previous Chapter, (1.1a) provides an example of a Volterra functional differential equation:

$$y'(x) = F(x, y(\ )) \tag{1.2}$$

in which the value of $F(x,y(\ )) \equiv F(x,y)$, at a point, depends on x and on $y(t)$ for $0 \leq t \leq x$; a number of the papers on delay differential equations cover the more general equation (1.2), and hence apply also to (1.1a). We shall employ $F(x,y)$ as a shorthand notation for the right-hand side of (1.1a), where convenient.

Problems of the type (1.1a,b) have practical interest, and in particular they occur in the study of competing populations (Davis 1962). Feldstein and Sopka (1973) list other areas of application. Pouzet

(1962) and Davis (1962) give some theoretical background for (1.1a,b); Linz (1969) gives sufficient conditions for the existence of a unique solution.

Numerical methods for the approximate solution of the initial-value problem (1.1a,b) have been presented in the literature, but the development of practical adaptive algorithms has been somewhat limited. For a fairly general study we refer to Feldstein and Sopka (1974). Published codes of algorithms for the treatment of various types of integro-differential equations, including†

$$y'(x) = \Phi(x,y(x), \int_0^x G(x, y(x), t, y(t), y'(t))\, dt),$$

with y(0) given, have been supplied by Pouzet (CNRS 1970) whose doctoral thesis includes a study of the problem. The methods of Pouzet can be applied, in particular, to (1.1a,b). We also observe that the algorithm of Neves (1975) can be applied to (1.1a,b), as he indicates.

2. NUMERICAL TECHNIQUES

Numerical methods for the approximate solution of the initial-value problem (1.1a,b) are generally derived by one of two apparently distinct types of approach. In fact, the resulting methods do not fall into two mutually exclusive classes.

Our first class of methods involves the direct adaptation of methods or of formulae for the initial value problem for

$$y'(x) = f(x,y) \equiv f(x, y(x)) \tag{2.1}$$

to the treatment of (1.2) (or, equivalently, of (1.1a)). Thus we can extend methods for the treatment of (2.1) to the treatment of the equation $y'(x) = f(x, y(x), v(x))$ where $v(x)$ is a suitable functional depending on $y(x)$. This is the usual approach adopted in the numerical solution of delay equations. Since the function $f(x,y,v)$ in (1.1a) has as one of its arguments the functional $v(x) = \int_0^x K(x,t, y(t))dt$, the adaptation of methods for (2.1) to treat (1.2) involves auxiliary approximations to $v(x)$. (This is analogous to the interpolation problem for a delay equat-

† Pouzet has also supplied an algorithm for the second-order equation analogous to (1.1a)

ion, see Chapter 20.) In the approximation of values of $v \equiv v(x)$, with

$$v(x) = \int_0^x K(x,t, y(t))\, dt,$$

it is usual to rely on the type of approximation employed in the numerical solution of Volterra integral equations (see for example, Kershaw 1974; Baker 1976). Methods which fall into our first class have been discussed by Linz (1969), Brunner and Lambert (1974), Tavernini (1971) and Neves (1975).

A second class of methods arise on writing $y' = F(x,y)$ as $y(x) = \int_0^x F(s, y(s))ds + y_0$ and applying integral equation techniques. In more detail, we have, for (1.1a,b),

$$\left.\begin{aligned} y(x) &= \int_0^x f(s,y(s), v(s))\, ds + y_0 \\ v(s) &= \int_0^s K(s,t,y(t))dt. \end{aligned}\right\} \tag{2.2}$$

These equations form a coupled pair of integral equations, which can be combined, using vector notation $\underset{\sim}{\phi}(x) = [y(x), v(x)]^T$ etc., into the form,

$$\underset{\sim}{\phi}(x) = \int_0^x \Phi(s, \underset{\sim}{\phi}(s))ds + \underset{\sim}{\phi}(0). \tag{2.3}$$

Techniques for the numerical solution of Volterra integral equations can readily be adapted to treat the system (2.3), but a somewhat wider range of formulae can be obtained if we treat each equation in (2.2) separately.

We shall, in the next two sections, introduce the reader to formulae for the approximate solution of (1.1a,b). We shall treat the two classes of methods indicated above, but it is not our intention to suggest that the two classes are mutually exlusive. Though the second class of methods appears to have been originally conceived (Mocarsky 1971) as distinct from the first class (Linz 1969), there is some overlap when natural simplifications are made to the computational scheme. (We show below that some of the methods of the second class reduce to methods of the first class. It may be remarked that any technique for a Volterra integral equation of the second kind provides *a* method for (2.1) on considering the integrated form: $y(x) = \int_0^x f(t,y(t))dt + y(0)$.)

3. METHODS OF THE FIRST CLASS

Simple methods like Euler's method or the trapezium rule method for (2.1) can readily be adapted to the treatment of (1.1a,b), given suitable methods for approximating $\int_0^{rh} K(rh, t, y(t))dt$. Thus, if we employ a simple equal-interval rule such as the trapezium rule, in composite form, we can adapt Euler's method to give the formula

$$y_{r+1} = y_r + hf(rh, y_r, h \sum_{j=0}^{r}{}'' K(rh, jh, y_j)) \qquad (3.1)$$

for $r \geq 0$, from which we can obtain, at the r-th stage, a value y_{r+1} approximating $y((r+1)h)$. For the trapezium rule method, we again use the repeated trapezium rule to approximate the integrals, giving the formula

$$y_{r+1} = y_r + \{\tfrac{1}{2}h\, f(rh, y_r, h \sum_{j=0}^{r}{}'' K(rh, jh, y_j)) + f((r+1)h, y_{r+1}, h \sum_{j=0}^{r+1}{}'' K((r+1)h, jh, y_j))\}. \qquad (3.2)$$

The simple formulae (3.1) and (3.2) are particular examples of methods based on multistep formulae.

A general class of linear multistep methods for (1.1) has been studied by Linz (1969). The general formula assumes the form:

$$\sum_{i=0}^{k} \alpha_i y_{n+i} = h \sum_{i=0}^{k} \beta_i \tilde{F}_{n+i}, \quad (n \geq 0) \qquad (3.3a)$$

where

$$\tilde{F}_r = f(rh, y_r, h \sum_{j=0}^{r} \omega_{rj} K(rh, jh, y_j)), \quad r \geq 0. \qquad (3.3b)$$

The formula is derived from the usual multistep formula for (2.1) together with a family of quadrature formulae $\int_0^{rh} \phi(s)ds \simeq h \sum_{j=0}^{r} \omega_{rj}\phi(jh)$. The formulae (3.3a,b) may be associated with the usual stability polynomials $\rho(z) = \sum_{i=0}^{k} \alpha_i z^i$, $\sigma(z) = \sum_{i=0}^{k} \beta_i z^i$, and the method is said to be zero-stable if the usual root-condition for zero-stability is satisfied. If, in addition to being zero-stable, the multistep formula for (2.1) is consistent, so that

$$\rho(1) = 0 \quad , \quad \sigma(1) = \rho'(1), \qquad (3.4a)$$

and if also

$$\limsup_{\substack{h\to 0\\ rh\leq X}} \left| \int_0^{rh} K(rh,s,y(s))ds - h \sum_{j=0}^{r} \omega_{rj} K(rh,jh,y(jh)) \right| = 0, \quad (3.4b)$$

then $\limsup_{\substack{h\to 0\\ jh\leq X}} |y(jh) - y_j| = 0$. In particular, given a suitable choice of $\rho(z)$ and $\sigma(z)$, convergence occurs if $K(x,s,y(s))$ is continuous for $0\leq s\leq x\leq X$ and the quadrature rules with weights $h\omega_{rj}$ are (Baker 1968; Davis and Rabinowitz 1975) Riemann sums. It is therefore possible to form the family of quadrature rules using composite versions of such simple rules as Simpson's rule, the trapezium rule, and the mid-point rule. The use of Riemann sums is sufficient (but not necessary) to ensure that for all h with $rh\leq X$

$$\left| \int_0^{rh} \phi(y)dy - h \sum_{j=0}^{r} \omega_{rj} \phi(jh) \right| \leq C\omega(\phi;h), \quad (3.5)$$

where $\omega(\phi;h)$ is the modulus of continuity of $\phi(x)$, and this guarantees (3.4b) when the function $K(x,s,y(s))$ is continuous.

The accuracy obtainable using (3.2) depends upon the local truncation error and the stability properties. The accuracy of the formulae associated with $\rho(z)$ and $\sigma(z)$ for the initial-value problems (2.1) has been discussed elsewhere, and we observe that a convenient family of quadrature rules with weights $h\omega_{rj}$ can be obtained using the Lagrangian form of Gregory's rule (Baker 1976) incorporating p-th differences. (If there is a fixed upper limit on p, independent of r, then (3.5) is satisfied.) Other quadrature formulae, based on Simpson's rule, will be demonstrated in anothe[r] context in Section 4.

As in the treatment of differential equations, and of Volterra integral equations of the second kind (Baker 1976), a convergence result is not sufficient to ensure a good computational scheme, and it is necessary to consider stability problems. A theory of stability regions and, in par[-]ticular, of A-stability for the method (3.3) has been given by Brunner an[d] Lambert (1974) who obtain stability regions by considering the scheme applied to the equation,

$$y'(x) = \xi y(x) + \eta \int_0^x y(t)dt. \quad (3.6)$$

The stability regions are functions of $h\xi$ and $h^2\eta$, and a number of specific results for simple formulae are illustrated by Brunner and Lambert. We give the following example, where shaded regions indicate the region of the parameters $h^2\eta, h\xi$, which yield stable relations.

$(\rho, \sigma) \leftrightarrow$ Backward Euler
$\{h\omega_{rj}\} \leftrightarrow$ Repeated trapezium rule

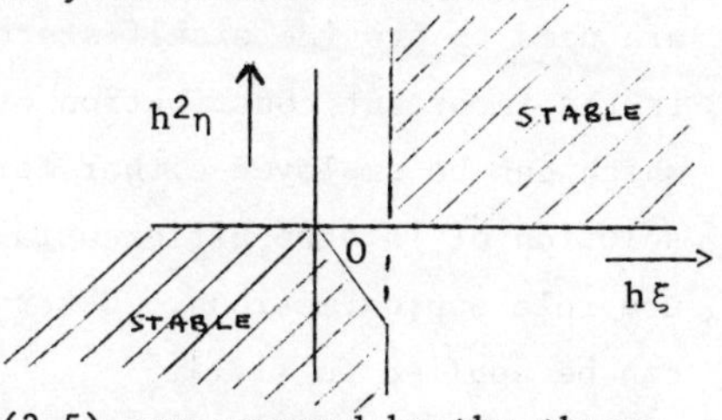

Not all quadrature formulae satisfying (3.5) are covered by the theory of Brunner and Lambert. The multistep methods mentioned above are not self-starting, and we consider the possibility of using self-starting 'one-step' methods. Linz (1969) considers some starting methods. We note, in passing, the possibility of employing Runge-Kutta type formulae (Pouzet 1970).

Feldstein and Sopka (1973) encounter a difficulty when they try to adapt the fourth-order Runge-Kutta formula, and the method they devise employs five evaluations of the function $f(x,y,v)$, rather than the four which might be expected from experience with differential equations, and is actually not self-starting. They employ $O(h^4)$ approximations $\tilde{J}(rh, \{y_s\}_0^r, \gamma, \delta)$ for $\int_0^{rh} K(rh+\gamma, s, y(s))ds + \int_{rh}^{rh+\gamma} K(rh+\gamma, s, y(rh)+(s-rh)\delta)ds$ which are obtained in terms of $y_0, y_1, y_2, \ldots, y_r$. (The last integral is an approximation, for suitable δ, to $\int_{rh}^{rh+\gamma} K(rh+\gamma, s, y(s))ds$). With such approximations, formed by weighted sums, the algorithm of Feldstein and Sopka assumes the form

$$
\begin{aligned}
L_0^{[r]} &= f(rh, y_r, \tilde{J}(rh, \{y_s\}_0^r, 0, 0)) \\
L_1^{[r]} &= f((r+\tfrac{1}{2})h, y_r + \tfrac{1}{2}hL_0^{[r]}, \tilde{J}(rh, \{y_s\}_0^r, \tfrac{1}{2}h, L_0^{[r]})) \\
L_2^{[r]} &= f((r+\tfrac{1}{2})h, y_r + \tfrac{1}{2}hL_1^{[r]}, \tilde{J}(rh, \{y_s\}_0^r, \tfrac{1}{2}h, L_0^{[r]})) \\
L_3^{[r]} &= f((r+\tfrac{1}{2})h, y_r + \tfrac{1}{2}hL_1^{[r]}, \tilde{J}(rh, \{y_s\}_0^r, \tfrac{1}{2}h, L_1^{[r]})) \\
L_4^{[r]} &= f((r+1)h, y_r + hL_3^{[r]}, \tilde{J}(rh, \{y_s\}_0^r, h, L_3^{[r]})) \\
y_{r+1} &= y_r + \tfrac{1}{6}h\left[L_0^{[r]} + 2L_1^{[r]} + 2L_2^{[r]} + L_4^{[r]}\right], \\
r &= 0, 1, \ldots, .
\end{aligned}
\tag{3.7}
$$

For further details of (3.7) we refer to the original work.

If the complexity of (3.7) suggests some difficulty in employing Runge-Kutta formulae, it is worth mentioning that in the work of Pouzet (1960, 1970), the usual Runge-Kutta formulae for differential equations are used to provide a self-starting method for (1.1). Also of interest is the important contribution of Tavernini (1971), who produces formulae which can be employed either for delay equations or for the numerical solution of integro-differential equations (1.1), given suitable quadrature rule approximations. Observe also that the approach of Neves (1975) can be applied to (1.1).

4. METHODS OF THE SECOND CLASS

We turn now to the application of integral equation techniques, to provide methods for (1.1). We consider the form (2.2) with $y(0) = y_0$.

The simplest example is provided by an adaption of the quadrature method. This involves the choice of quadrature formulae

$$\int_0^{rh} \phi(s)ds \simeq h \sum_{j=0}^{r} \omega_{rj}\phi(jh), \quad \int_0^{rh} \psi(s)ds \simeq h \sum_{j=0}^{r} \omega'_{rj}\psi(jh) \tag{4.1}$$

which are employed to discretize (2.2). We obtain the system of equations

$$\left.\begin{aligned} y_r &= h \sum_{j=0}^{r} \omega'_{rj} f(jh, y_j, v_j) + y_0 \quad (r = 1,2, ,..,) \\ v_j &= h \sum_{i=0}^{j} \omega_{ji} K(jh, ih, y_i) \quad (j = 1,2,...r); \end{aligned}\right\} \tag{4.2}$$

See Mocarsky (1971).

The solution of (4.2) proceeds in a step-by-step fashion, and generally involves rather more summation than the use of (3.3a), but no additional storage requirement. As we observe below, the work involved with the use of (4.2) can generally be reduced by rearranging the equations, thus reducing the method to one based on the form (3.3a,b) or a similar type of equation. Where such a rearrangement is possible, it should certainly be executed. The method of (4.2) can be shown to be convergent, as $h \to 0$, if the two rules in (4.1) (for $r = 1,2,...$) satisfy the condition of (3.5). See Mocarsky (1971 for this result.

Various choices of weights ω_{rj}, ω'_{rj} employed in the numerical solution of Volterra integral equations (Baker 1976) can be used in (4.2). To simplify, we may suppose that $\omega_{rj} = \omega'_{rj}$. As a simple example, suppose

$\omega_{rj} = \omega'_{rj}$, $j=0,1,\ldots,r$, with $\omega_{rj} = 1$ for $j = 1\ 2,\ldots,r-1$ and $\omega_{r0} = \omega_{r1} = \frac{1}{2}$. This corresponds to the use of repeated trapezium rules. Differencing the resulting equations for y_r in (4.2) we reduce the method to the form (3.2); thus, for $r = 0,1,\ldots$ we have

$$y_{r+2} - y_r = \tfrac{1}{2}h(f_r+f_{r+1}),\ f_j = f(jh,\ y_j,\ v_j),\ v_j = h \sum_{i=0}^{j}{}'' K(jh,\ ih,\ y_i).$$

If $K(x,t,y)$ is independent of x, v_j can be obtained from v_{j-1}.

The repeated trapezium rule is a 'trivial' Gregory rule. In addition to the Gregory rules, mentioned in Section 3, we may employ the repeated Simpson's rule, combined with the single step trapezium rule to give either,

$$j = 0,\ 1,\ 2,\ \ldots$$

$$\omega'_{rj} = \left\{\begin{array}{ll} 1/3,\ 4/3,\ 2/3,\ \ldots,\ 4/3,\ 1/3 & (r \text{ even}) \\ 1/3,\ 4/3,\ 2/3,\ \ldots,\ 4/3,\ 5/6,\ 1/3 & (r \text{ odd},\ r > 1 \end{array}\right\} \tag{4.3}$$

or

$$\omega'_{rj} = \left\{\begin{array}{ll} 1/3,\ 4/3,\ 2/3,\ \ldots,\ 4/3,\ 1/3\ (r \text{ even}) \\ 1/2,\ 5/6,\ 4/3,\ \ldots,\ 2/3,\ 4/3,\ 1/3 & (r \text{ odd},\ r > 1) \end{array}\right\}, \tag{4.4}$$

for example. (For $r = 1$ we may set $\omega'_{11} = \omega'_{10} = 1/2$, which amounts to using the trapezium rule as a starting procedure.)

Whilst the local truncation errors of the formula (4.2), constructed using (4.3) or (4.4) for the weights $\omega_{rj} = \omega'_{rj}$, is $O(h^3)$ in both cases, the practical use of (4.4) is not to be recommended because of instability. Stability regions have been computed, at Manchester, for the formula (4.3) and (4.4) applied to equation (3.6), thus developing a theory proposed for integral equations by Baker (1976). Regions of stability for (4.3) and (4.4) are indicated below, by the 'shaded' areas:

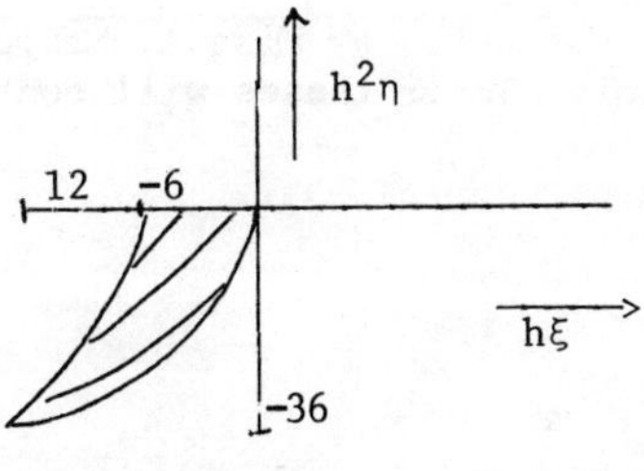

(4.3)

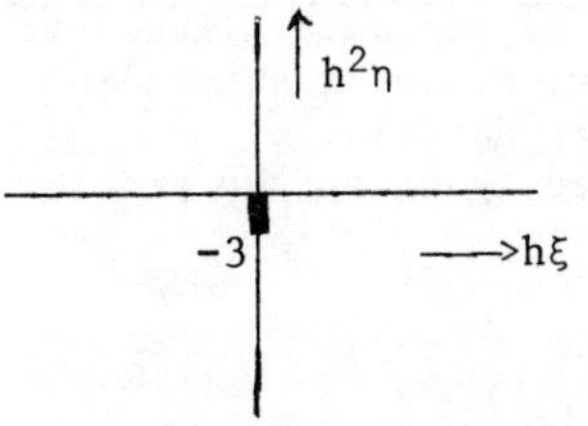

(4.4)

For the stability of (4.4) we require $\xi = 0$, $\eta < 0$ and $h^2\eta > -3$. (The relationship to the stability region for the Simpson's rule corrector formula for $y' = \lambda y$, requiring λ imaginary and $|\lambda| < \sqrt{3}$ may be noted.) More details of these and related results for other formulae will be published elsewhere†. To obtain insight into the stability result for (4.4) we can difference the equations for y_r in (4.2), thereby obtaining $y_{r+2} - y_r = \frac{1}{3}h\{f_{r+2}+4f_{r+1}+f_r\}$ with $f_j = f(jh,y_j,v_j)$ for $j = 0,1,2, \ldots$, and showing the equivalence of the method (4.2) using (4.4), with (3.3a) using the weakly-unstable Simpson's formula.

Whilst the preceding remark shows the reason for the instability with the choice (4.4), observe that differencing the equations for the successive values y_r in (4.2) frequently leads to a computationally more effective scheme, even if it does <u>not</u> lead directly to a multistep formula (3.3a,b). Examples where (4.2) proves equivalent to a multistep formula (3.3) are provided by choosing the weights ω'_{rj} to correspond to the Gregory rules with a fixed number of differences*: subtracting the equation for y_r from that for y_{r+1} in (4.2), we obtain an Adams corrector formula. (We have already seen that the trapezium rule with $\omega'_{rj} = 1$ if $j \notin \{0,r\}$, $\omega'_{rr} = \omega'_{ro} = \frac{1}{2}$, gives the trapezium rule (3.3a) with $y_{r+1}-y_r = \frac{1}{2}h\{f_r+f_{r+1}\}$ where $f_j = f(jh,y_j,v_j)$ as before.) An attempt to reduce (4.4) to a multistep formula when using the weights (4.3) fails; instead we produce the cyclic equations

$$y_{2r+1} - y_{2r} = \tfrac{1}{2}h\{f_{2r+1} + f_{2r}\}$$

$$y_{2r} - y_{2(r-1)} = \frac{1}{3}h\{f_{2r} + 4f_{2r-1} + f_{2r-2}\};$$

which, however, again prove compuationally more convenient than the use of the first equation in (4.2).

High-order quadratures methods of the form (4.2) generally require starting procedures. A self-starting procedure which could be employed as a starting method giving $O(h^4)$ accuracy has been supplied by Day (1970)

†Some of the work alluded to here has been studied by Miss A. Makroglou and Mr. E. Short as students of the author. Their theses will contain further details.

*We are ignoring the problem of starting procedures here.

5. BLOCK-BY-BLOCK METHODS

Block-by-block implicit methods developed for differential equations and for the approximate solution of Volterra integral equations can be extended to allow treatment of (1.1a,b). We shall again study the form (2.2) and employ the approach suggested by integral equations. The required formulae are based upon the use of a basic quadrature rule:

$$\int_0^1 \phi(t)\,dt \simeq \sum_{j=0}^{m} \omega_j \phi(\theta_j) \tag{5.1}$$

where

$$0 \leqslant \theta_0 < \theta_1 < \cdots < \theta_{m-1} < \theta_m \leqslant 1.$$

We also require further integration formulae, of the form¶

$$\int_0^{\theta_r} \phi(t)\,dt \simeq \sum_{j=0}^{m} \omega_{rj}\, \phi(\theta_j) \quad (r=0,1,\ldots,m). \tag{5.2}$$

Such formulae as (5.1) and (5.2) can be obtained by integrating the polynomial interpolating $\phi(t)$ at $\theta_0, \theta_1 \ldots, \theta_m$ over $[0,1]$ or $[0,\theta_r]$ respectively. (With $\ell_k(t) = \prod_{i\neq k}\{(t-\theta_i)/(\theta_k-\theta_i)\}$ we then obtain $\omega_j = \int_0^1 \ell_j(t)dt$, $\omega_{rj} = \int_0^{\theta_r} \ell_j(t)dt$. A different choice for (5.2) is mentioned below.

The use of the quadrature forulae (5.1) and (5.2), adapted to the range of integration, enables us to discretize the integrals in (2.2) for $x \in \{x_{rs}\}$ where $x_{rs} = rh + \theta_s h$. We obtain, using this notation,

$$y_{rs} = h \sum_{k=0}^{r-1} \sum_{j=0}^{m} \omega_j f(x_{kj}, y_{kj}, v_{kj}) + h \sum_{j=0}^{m} \omega_{sj} f(x_{rj}, y_{rj}, v_{rj}) + y_0 \tag{5.3}$$

$$v_{kj} = h \sum_{i=0}^{k-1} \sum_{p=0}^{m} \omega_p K(x_{kj}, x_{ip}, y_{ip}) + h \sum_{p=0}^{m} \omega_{jp}\, K(x_{kj}, x_{kp}, y_{kp}). \tag{5.4}$$

As with the quadrature method (4.2), the first of these equations, that is equation (5.3), can be simplified. We effectively reduce the method to a method of the first class, obtained by discretizing the equation $y(x_{rs}) - y(x_{ro}) = \int_{x_{ro}}^{x_{rs}} y'(t)dt$, using a block-by-block method. Thus we obtain from equation (5.3) (on differencing and supposing that $\theta_0 = 0$ with $\omega_{oj} = 0$) the equation

¶ If $\theta_m=1$, then we suppose that $\omega_{mj}=w_j\,(j=0,1,\ldots,m)$.

$$y_{rs} - y_{r0} = h \sum_{j=0}^{m} \omega_{sj} f(x_{rj}, y_{rj}, v_{rj}), \quad (s=0,1,\ldots,m;\ r=0,1,\ldots) \tag{5.5}$$

If $\theta_m = 1$, then $y_{rm} = y_{r+1,o}$, so that equation (5.5) is all that is required, along with the equation (5.4) for v_k. Obvious modifications are required if $\theta_0 \neq 0$, $\theta_m \neq 1$. A popular choice of $\{\theta_i\}$, in which $\theta_0 = 0$ and $\theta_m = 1$, yields the Lobatto quadrature rule in (5.1).

Some variations on the above approach are possible; it is reasonable, for example, to replace our suggested construction of (5.2) by the combination of (5.1) with polynomial interpolation. Thus, when $\theta_0 = 0$ and $\theta_m = 1$, we can replace (5.5) and (5.4) by the equations $y_{r+1,s} = y_{r,m}$, and

$$y_{rs} - y_{r0} = \theta_s h \sum_{j=0}^{m} \omega_j f(x_{sj}, \tilde{y}((\theta_s\theta_j + r)h),\ \tilde{v}((\theta_s\theta_j + r)h))$$

where

$$v_{kj} = h \sum_{i=0}^{k-1} \sum_{p=0}^{m} \omega_p K(x_{kj}, x_{ip}, y_{ip}) + \theta_j h \sum_{p=0}^{m} \omega_p K(x_{kj}, (\theta_j\theta_p + k)h, \tilde{y}((\theta_j\theta_p + k)h))$$

and where

$$\tilde{y}((\theta_s\theta_j + r)h) = \sum_{k=0}^{m} \{\prod_{i \neq k} \frac{\theta_s\theta_j - \theta_i}{\theta_k - \theta_i}\} y((\theta_k + r)h),$$

and we can set $$\tilde{v}((\theta_s\theta_j + r)h) = \sum_{k=0}^{m} \{\prod_{i \neq k} \frac{\theta_s\theta_j - \theta_i}{\theta_k - \theta_i}\} v((\theta_k + r)h) .$$

(In this way, we avoid use of values $f(x, \tilde{y}(s), \tilde{v}(s))$ and $K(x, s, \tilde{y}(s))$ in which s exceeds x in value.)

Methods based on (5.3) and (5.4) or variants, generally involve considerable work per step, but with appropriate choice of $\{\theta_i\}$ should have stability properties which permit a large step-size h in appropriate cases.

6. TREATMENT OF WEAK SINGULARITIES

Product integration formulae can be employed to deal with the presence of weak singularities, for example when

$$K(x,t, y(t)) = G(x,t, y(t))/(x-t)^{\alpha}, \tag{6.1}$$

where $0 \leqslant \alpha < 1$ and $G(x,t, y(t))$ is continuous for $0 \leqslant t \leqslant x \leqslant X$, in (1.1a). In such circumstances, the sums previously suggested as a discretization for the integral term in (1.1a) are generally infinite, but our methods can be modified by using appropriate product integration formulae.

Thus, for (6.1) we might employ the generalized trapezium rule giving approximations of the form

$$\int_0^{rh} \{\phi(t)/(rh-t)^{\alpha}\}dt \simeq \sum_{j=0}^{r} \nu_{rj}\phi(jh), \quad (r=1,2,\ldots), \tag{6.2}$$

where the weights ν_{rj} are chosen so that (6.2) is exact if $\phi(t)$ is continuous, and linear in each interval $[0,h]$, $[h,2h]$, ..., $[(r-1)h, rh]$. More generally, we can adapt the type of schemes illustrated by (4.3) and (4.4) to product integration formulae; the range of possibilities also includes product-integration analogues of (5.3) and (5.4). Details of such techniques for approximating weakly singular integrals can be found in the literature on weakly singular Volterra integral equations (in particular, in Baker 1977 §6.8).

REFERENCES

ABRAMOWITZ, M. and STEGUN, I.A. (1964) Handbook of Mathematical Functions. National Bureau of Standards, Washington.

ALLEN, R.H. (1969) Numerically stable explicit integration techniques using a linearized Runge-Kutta extension. Joint Conference on Mathematical and Computer Aids to Design.

ALSPAUGH, D.W. (1974) The application of invariant imbedding to the solution of linear two-point boundary value problems on an infinite interval. Math. Comp. 28, 1005.

ALSPAUGH, D.W., KAGIWADA, H.H. and KALABA, R. (1970) Application of invariant imbedding to the buckling of columns. J. Comp. Phys. 5, 56.

ARIS, R. (1969) Elementary Chemical Reactor Analysis. Prentice-Hall, Englewood Cliffs, N.J.

AVILA, J.H. (1974) The feasibility of continuation methods for non-linear equations. SIAM J. Numer. Anal. 11, 102.

AXELSSON, O. (1969) A class of A-stable methods. BIT. 9, 185-199.

AXELSSON, O. (1972) A note on a class of strongly A-stable methods. BIT. 12, 1-4.

AZIZ, A.K. (ed.) (1975) Numerical Solution of Boundary Value Problems for Ordinary Differential Equations. Academic Press.

BABUSKA, I., PRAGER, M. and VITASEK, E. (1966) Numerical Processes in Differential Equations. Wiley (Interscience) New York.

BAILEY, P.B. (1966) Sturm-Liouville eigenvalues via a phase function. J. SIAM. Appl. Math. 14, 242.

BAKER, C.T.H. (1968) On the nature of certain quadrature formulae and their errors. SIAM J. Numer. Anal. 5, 783-804.

BAKER, C.T.H. (1974) Methods for integro-differential equations, in DELVES and WALSH (1974). 189-206.

BAKER, C.T.H. (1977) The Numerical Treatment of Integral Equations. Clarendon Press: Oxford.

BALL, S.J. (1966) A variable time-delay subroutine for digital simulation programs. Simulation. 7 (1).

BANKS, D.O. and KUROWSKI, G.J. (1968) Computation of eigenvalues of singular Sturm-Liouville systems. Math. Comp. 22, 304.

BANKS, D.O. and KUROWSKI, G.J. (1973) Computation of eigenvalues for vibrating beams by use of Prüfer transformation. SIAM J. Numer. Anal. 10, 918.

BARRODALE, I. and YOUNG, A. (1970) Computational experience in solving linear operator equations using the Chebyshev norm, in Numerical Approximations to Functions and Data, ed. J.G. Hayes, Athlone Press. 115-142.

BARTON, D., WILLERS, I.M. and ZAHAR, R.V.M. (1971) Taylor series methods for ordinary differential equations - an evaluation, from Mathematical Software, ed. J.R. Rice, Academic Press.

BARWELL, V.K. (1974) On the asymptotic behaviour of the solution of a differential-difference equation. Utilitas Mathematica. 6, 189-194.

BARWELL, V.K. (1975) Special stability problems for functional differential equations. BIT. 15, 130-135.

BAUER, F.L. (1963) Optimally scaled matrices. Num. Math. 5, 73-87.

BELLMAN, R.E. (1961) On the computational solution of differential-difference equations. J. Math. Anal. and its Appl. 2, 108-110.

BELLMAN, R.E. (ed.) (1962) Mathematical Problems in the Biological Sciences. Proc. Symp. Appl. Math. XIV., American Mathematical Society, Providence, R.I.

BELLMAN, R.E., BUELL, J.D. and KALABA, R.E. (1965) Numerical integration of a differential-difference equation with a decreasing time-lag. Communs. Ass. comput. Mach. 8, 227-228.

BELLMAN, R.E., BUELL, J.D. and KALABA, R.E. (1966) Mathematical experimentation in time-lag modulation. Communs. Ass. comput. Mach. 9, 752-754.

BELLMAN, R.E. and COOKE, K.L. (1963) Differential-difference Equations.

Academic Press, New York.

BELLMAN, R.E. and COOKE, K.L. (1965) On the computational solution of a class of functional differential equations. J. Math. Anal. and its Appl. 12, 495-500.

BELLMAN, R.E. and KOTKIN, B. (1962) On the numerical solution of a differential-difference equation arising in analytic number theory. Math. Comp. 16, 473-475.

BEREZIN, I.S. and ZHIDOV, N.P. (1965) Computing Methods Volumn II. Pergamon Press.

BJÖRCK, A. (1967) Solving linear least squares problems by Gram-Schmidt orthogonalisation. BIT. 7, 1.

BJUREL, G., DAHLQUIST, G., LINDBERG, B., LINDE, S. and ODEN, L. (1970) Survey of stiff ordinary differential equations. Rept. NA. 70.11, Dept. of Inf. Proc., Royal Inst. of Tech., Stockholm.

BRAYTON, R.K. and WILLOUGHBY, R.A. (1967) On the numerical integration of a symmetric system of difference-differential equations of neutral type. J. Math. Anal. and its Appl. 18, 182-189.

BRAYTON, R.K., GUSTAVSON, F.G. and HACHTEL, G.D. (1972) A new efficient algorithm for solving differential-algebraic systems using implicit backward differentiation formulas. Proc. IEEE. 60, 98-108.

BRUNNER, H. (1972) A class of A-stable two-step methods based on Schur polynomials. BIT. 12, 468-474.

BRUNNER, H. (1973) On the numerical solution of nonlinear Volterra integro-differential equations. BIT. 13, 381-390.

BRUNNER, H. and LAMBERT, J.D. (1974) Stability of numerical methods for Volterra integro-differential equations. Computing. 12, 75-89.

BULIRSCH, R. and STOER, J. (1964) Fehlerab Schatzungen und Extrapolation mit Rationalen Funletionen bei Verfahren vom Richardson-Typus. Num. Math. 6, 413-427.

BULIRSCH, R. and STOER, J. (1966) Numerical treatment of ordinary differential equations by extrapolation methods. Num. Math. 8, 1-13.

BUSINGER, P. and GOLUB, G.H. (1965) Linear least squares solutions by Householder transformations. Num. Math. 7, 269.

BUTCHER, J.C. (1963) Coefficients for the study of Runge-Kutta integration processes. J. Austral. Math. Soc. 3, 185-201.

BUTCHER, J.C. (1964) Implicit Runge-Kutta processes. Math. Comp. 18, 50-64.

BUTCHER, J.C. (1965a) On the attainable order of Runge-Kutta methods. Math. Comp. 19, 408-417.

BUTCHER, J.C. (1965b) A modified multistep method for the numerical integration of ordinary differential equations. J. Ass. comput. Mach. 12, 124-135.

BUTCHER, J.C. (1966) On the convergence of numerical solutions to ordinary differential equations. Math. Comp. 20, 1-10.

BUTCHER, J.C. (1969) The effective order of Runge-Kutta methods, from Lecture Notes in Mathematics. 109, Springer. 133-139.

BYRNE, G.D. and HINDMARSH, A.C. (1975) A polyalgorithm for the numerical solution of ordinary differential equations. ACM Trans. on Math. Software. 1, 71-96.

CALAHAN, D.A. (1967) Numerical solution of linear systems with widely separated time constants. Proc. IEEE. 55, 2016-2017.

CEBICI, T. and KELLER, H.B. (1971) Shooting and parallel shooting methods for solving the Falkner-Skan boundary layer equations. J. Comp. Phys. 7, 289.

CESCHINO, F. and KUNTZMAN, J. (1963) Numerical Solution of Initial Value Problems. Dunod Davis; English Translation (1966) Prentice Hall.

CHARTRES, B. and STEPLEMAN, R. (1971) Convergence of difference methods for initial and boundary value problems with discontinuous data. Math. Comp. 24, 729-732.

CHARTRES, B. and STEPLEMAN, R. (1972) A general theory of convergence for numerical methods. SIAM. J. Numer. Anal. 9, 476-492.

CHIPMAN, F.H. (1971a) A-stable Runge-Kutta processes. BIT. 11, 384-388.

CHIPMAN, F.H. (1971b) Numerical Solution of Initial Value Problems using A-stable Runge-Kutta Processes. Thesis, Dept. of AACS, University of Waterloo.

CHOSKY, N.H. (1960) Time-lag systems - Bibliography IRE Trans. Automatic Control. AC-5.

C.N.R.S. (1970) Procédures ALGOL en Analyse Numerique. Centre National de la Recherche Scientifique, Paris.

CODDINGTON, E. and LEVINSON, N. (1955) Theory of Ordinary Differential Equations. McGraw-Hill, New York.

COLE, J.D. (1968) Perturbation Methods in Applied Mathematics. Blaisdell, Waltham, Mass.

CONTE, S.D. (1966) The numerical solution of linear boundary value problems. SIAM Rev. 8, 309.

COOKE, K.L. and LIST, S.E. (1972) The numerical solution of integro-differential equations with retardation. Tech. Rept. No. 72-4, Dept. of Elec. Eng., University of Southern California, Los Angeles.

CRAIGIE, J.A.I. (1975) A variable order multistep method for the numerical solution of stiff systems of ordinary differential equations. Numer. Anal. Rept. No. 11, Dept. of Math., University of Manchester.

CRANE, R.L. and KLOPFENSTEIN, R.W. (1965) A predictor-corrector algorithm with an increased range of absolute stability. J. Ass. Comput. Mach. 12, 227-241.

CRYER, C.W. (1972) Numerical methods for functional differential equations, in SCHMITT (1972). 17-102.

CRYER, C.W. (1973) A new class of highly-stable methods: A_0-stable methods. BIT. 13, 153-159.

CRYER, C.W. (1974) Highly stable numerical methods for delay differential equations. SIAM J. Numer. Anal. 11, 788-797. Also available in extended version as C.S. Tech. Rept. 190, University of Wisconsin at Madison (1973).

CRYER, C.W. and TAVERNINI, L. (1972) The numerical solution of Volterra functional differential equations by Euler's method. SIAM J. Numer. Anal. 9, 105-129; C.S. Tech. Rept. 71, University of Wisconsin at Madison (1969).

CURTIS, A.R., POWELL, M.J.D. and REID, J.K. (1974) On the estimation of sparse Jacobian matrices. J. Inst. Math. and its Appl. 13, 117-119.

CURTIS, A.R. and REID, J.K. (1974) The choice of step lengths when using differences to approximate Jacobian matrices. J. Inst. Math. and its Appl. 13, 121-126.

CURTISS, C.F. and HIRSCHFELDER, J.O. (1952) Integration of stiff equations. Proc. Nat. Acad. Sci., USA. 38, 235-243.

DAHLQUIST, G. (1956) Convergence and stability in the numerical integration of ordinary differential equations. Math. Scand. 4, 33-53.

DAHLQUIST, G. (1959) Stability and error bounds in the numerical integration of ordinary differential equations. Trans. of the Royal Inst. of Tech., Stockholm. 130.

DAHLQUIST, G. (1963) A special stability problem for linear multistep methods. BIT. 3, 27-43.

DAHLQUIST, G. (1969) A numerical method for some ordinary differential equations with large Lipschitz constants, in Information Processing 68 (Proceedings of the IF.P Congress 1968), North Holland Publishing Co.

DAHLQUIST, G. (1975) Dundee biennial numer. anal. conference. To be published by Springer-Verlag.

DAHLQUIST, G. and LINDBERG, B. (1973) On some implicit one-step methods for stiff differential equations. TRITA-NA-7302. Dept. of Inf. Proc., Royal Inst. of Tech., Stockholm.

DAVEY, A. (1973) A simple numerical method for solving Orr-Sommerfield problems. Q.J. Mech. Appl. Math. 26, 401.

DAVIS, H.T. (1962) Introduction to Nonlinear Differential and Integral Equations. Dover, New York.

DAVIS, P.J. (1963) Interpolation and Approximation. Blaisdell.

DAVIS, P.J. and RABINOWITZ, P. (1975) Methods of Numerical Integration. Academic Press, New York.

DAY, J.T. (1967) Note on the numerical solution of integro-differential equations. Comput. J. 9, 394-395.

DAY, J.T. (1970) On the numerical solution of integro-differential equations. BIT. 10, 511-514.

DELVES, L.M. and WALSH, J. (ed.) (1974) Numerical Solution of Integral Equations. Clarendon Press, Oxford.

DE NEVERS, K. and SCHMITT, K. (1971) An application of the shooting method to boundary-value problems for second-order delay equations. J. Math. Anal. and its Appl. 36, 588-597.

DEUFLHARD, P. (1974) A modified Newton method for the solution of ill-conditioned systems of non-linear equations with application to multiple shooting. Num. Math. 22, 289.

DILL, C. and GEAR, C.W. (1971) A graphical search for stiffly stable methods for ordinary differential equations. J. Ass. Comp. Mach. 18, 75-79.

DOUGLAS, J. and DUPONT, T. (1973) A finite element collocation method for quasilinear parabolic equations. Math. Comp. 27, 17-28.

DOUGLAS, J. and DUPONT, T. (1974) Collocation Methods for Parabolic Equations in a Single Space Variable. Springer-Verlag.

DRIVER, R.D. (1962) Existence and stability of solutions of a delay-differential system. Arch. Rat. Mech. Anal. 10, 401-426.

EDSBERG, L. (1974) Integration package for universal kinetics, in WILLOUGHBY (1974).

EHLE, B.L. (1968) High-order A-stable methods for the numerical solution of systems of differential equations. BIT. 8, 276-278.

EHLE, B.L. (1969a) On Padé approximations to the exponential function and A-stable methods for the numerical solution of initial-value problems. Res. Rept. CSRR 2010. University of Waterloo, also (1969b) Thesis, Dept. of AACS, University of Waterloo.

EHLE, B.L. (1972) A comparison of numerical methods for solving certain stiff ordinary differential equations. Tech. Rept. No. 70, Dept. of Math. University of Victoria.

EL-GENDI, S.E. (1969) Chebyshev solution of differential, integral and integro-differential equations. Comput. J. 12, 282-287.

EL'SGOL'TS, L.E. (1953) Approximate methods of integration of differential-difference equations. (Russian) Uspekhi Mat. Nauk. 8 (4).

El'SGOL'TS, L.E. and NORKIN, S.B. (1973) Introduction to the Theory and Applications of Differential Equations with Deviating Arguments. Academic Press, New York.

ENGLAND, R. (1957) Automatic Methods for Solving Systems of Ordinary Differential Equations. Ph.D. Thesis, University of Liverpool.

ENGLAND, R. (1959) Error estimates for Runge-Kutta type solutions to systems of O.D.E.'s. Comput. J. 12, 166-169.

ENRIGHT, W.H. (1974) Second derivative multistep methods for stiff ordinary differential equations. SIAM J. Numer. Anal. 11, 321-331.

ENRIGHT, W.H., BEDET, R., FARKAS, I. and HULL, T.E. (1974) Test results on initial value methods for non-stiff ordinary differential equations. Tech. Rept. No. 68. Dept. of Comput. Sci., University of Toronto.

ENRIGHT, W.H., HULL, T.E., and LINDBERG, B. (1974) Comparing numerical methods for stiff systems of O.D.E.'s. Tech. Rept. No. 69.

Dept. of Comput. Sci., University of Toronto.

FEHLBERG, E. (1968) Classical fifth, sixth, seventh and eight order Runge-Kutta formulas with stepsize control. NASA. T.R. 287.

FEHLBERG, E. (1969a) Klassische Runge-Kutta-Formeln funfter und siebenter Ordnung mit Schrittweiten-Kontrolle. Computing. 4, 93-106.

FEHLBERG, E. (1969b) Low order classical Runge-Kutta formulas with stepsize control and their application to some heat transfer problems. NASA. T.R. 315.

FELDSTEIN, A. (1964) Discretization Methods for Retarded Ordinary Differential Equations. Ph.D. thesis, University of California at Los Angeles.

FELDSTEIN, A. and GOODMAN, R. (1973) Numerical solution of ordinary and retarded differential equations with discontinuous derivatives. Num. Math. 21, 1-13.

FELDSTEIN, A. and SOPKA, J.R. (1974) Numerical methods for nonlinear Volterra integro-differential equations. SIAM. J. Numer. Anal. 11, 826-846.

FOX, L., MAYERS, D.F., OCKENDON, J.R. and TAYLOR, A.B. (1971) On a functional differential equation. J. Inst. Math. and its Appl. 8, 271-307.

FOX, L. and PARKER, I. (1968) Chebyshev Polynomials in Numerical Analysis. Oxford University Press.

FOX, P.A. (1971) DESUB: Integration of a first-order system of ordinary differential equations, in Mathematical Software, ed. J.R. Rice, Academic Press.

FREEMAN, L. and DELVES, L.M. (1974) On the convergence rates of variational methods. J. Inst. Math. and its Appl. 14, 311-323.

GEAR, C.W. (1964) Hybrid methods for the initial value problem in ordinary differential equations. SIAM. J. Numer. Anal. 2, 69-86.

GEAR, C.W. (1969) The automatic integration of stiff ordinary differential equations, in Information Processing 68 (Proceedings of the IFIP Congress 1968), North Holland Publishing Co. 187-193.

GEAR, C.W. (1971a) Algorithm 407, DIFSUB for solution of ordinary differential equations. Communs. Ass. comput. Mach. 14, 185-190.

GEAR, C.W. (1971b) Numerical Initial Value Problems in Ordinary

Differential Equations. Prentice Hall, Englewood Cliffs, New Jersey.
GEAR, C.W. (1971c) Simultaneous numerical solution of differential-algebraic equations. IEEE. Trans. on Circuit Theory. 18, 89-95.
GEAR, C.W. and TU., K.W. (1974) The effect of variable mesh size on the stability of multistep methods. SIAM. J. Numer. Anal. 11, 1025-1043.
GEAR, C.W. and WATANABE, D.S. (1974) Stability and convergence of variable order multistep methods. SIAM. J. Numer. Anal. 11, 1044-1058.
GEORGE, J.H. and GUNDERSON, R.W. (1972) Conditioning of linear boundary value problems. BIT. 12, 172.
GERSTING, J.M. Jr. and JANKOWSKI, D.F. (1972) Numerical methods for Orr-Sommerfeld problems. Int. J. for Num. Math. in Eng. 4, 195.
GODART, M. (1966) An iterative method for the solution of eigenvalue problems. Math. Comp. 20, 399.
GOODMAN, R. (1971) Round-off Error in the Numerical Solution of Retarded Ordinary Differential Equations. Ph.D. thesis, Harvard University, Cambridge, Mass.
GOODMAN, R. and FELDSTEIN, A. (1973) Round-off error for retarded ordinary differential equations: a priori bounds and estimates. Num. Math. 21, 355-372.
GORBUNOV, A.D. and BUDAK, B.M. (1958) The convergence of certain finite difference processes for equations $y' = f(x, y)$ and $y'(x) = f(x, y(x), y(x - \tau(x))$. Doklady Akad. Nauk. SSSR. 119 (4).
GOURLAY, A.R. (1970) A note on trapezoidal methods for the solution of initial value problems. Math. Comp. 24, 629-633.
GRAGG, W.B. (1965) On extrapolation algorithms for ordinary initial value problems. J. SIAM. Numer. Anal. 2, 384-403.
GRAGG, W.B. and STETTER, H.J. (1964) Generalized mutlistep predictor-corrector methods. J. Ass. comput. Mach. 11, 189-209.
GUDERLEY, K.G. (1973) A unified view of some methods for stiff two-point boundary value problems. ARL Tech. Rept. Aerospace Research Labs., Wright-Patterson Airforce Base. Ohio 45433.
GUDERLEY, K.G. and NICOLAI, P.J. (1966) Reduction of two-point boundary value problems in a vector space to initial value problems by projection. Num. Math. 8, 270.

HABETS, P. (1974) A consistency theory of singular perturbations of differential equations. SIAM J. Appl. Math. 26, 136-153.

HAHN, W. (1967) Stability of Motion, tr. by A.P. Baartz. Springer-verlag.

HAINES, C.F. (1969) Implicit integration processes with error estimates for the numerical solution of diffe ential equations. Comput. J. 12, 183-187.

HALANAY, A. (1966) Differential Equations - Stability, Oscillations, Time Lags. Academic Press, New York.

HALE, J. (1971) Functional Differential Equations. Springer, New York.

HALL, G. (1974) Stability analysis of predictor-corrector algorithms of Adams type. SIAM. J. Numer. Anal. 11, 494-505.

HALL, G., ENRIGHT, W.H., HULL, T.E., SEDGWICK, A.E. (1973) Detest: a program for comparing numerical methods for ordinary differential equations. Tech. Rept. No. 60. Dept. of Comput. Sci., University of Toronto.

HAMMER, P.C. and HOLLINGSWORTH, J.W. (1955) Trapezoidal methods of approximating solutions of differential equations. M.T.A.C. 9, 92-96.

HAMMING, R.W. (1959) Stable predictor-corrector methods for ordinary differential equations. J. Ass. comput. Mach. 6, 37-47.

HENRICI, P. (1962) Discrete Variable Methods in Ordinary Differential Equations. Wiley.

HILL, D.R. (1973) An Approach to the Numerical Solution of Delay-Differential Equations. Ph.D. thesis, University of Pittsburgh.

HILL, D.R. (1974) A new class of one-step methods for the solution of Volterra functional differential equations. BIT. 14, 298-305.

HOPKINS, T.R. (1975) On the Numerical Solution of Stiff Parabolic Partial Differential Equations. University of Liverpool. (to appear.)

HOPPENSTEADT, F. (1971) Properties of solutions of ordinary differential equations with small parameters. Communs. Pure and Appl. Math. 24, 807-840.

van der HOUWEN, P.J. and VERWER, J.G. (1974) Generalized linear multistep methods I: development of algorithms with zero-parasitic roots. Tech. Rept. NW 10/74, Mathematisch Centrum, Amsterdam.

HULL, T.E. (1974) A proposal for the design of subroutines to solve initial value problems associated with ordinary differential equations. Tech. Note.

HULL, T.E. and CREEMER, A.L. (1963) Efficiency of predictor-corrector schemes. J. Ass. comput. Mach. 10, 291-301.

HULL, T.E. and ENRIGHT, W.H. (1974) A structure for programs that solve ordinary differential equations. Tech. Rept. No. 66, Dept. of Comput. Sci., University of Toronto.

HULL, T.E., ENRIGHT, W.H., FELLEN, B.M. and SEDGWICK, A.E. (1972) Comparing numerical methods for ordinary differential equations. SIAM. J. Numer. Anal. 9, 603-637.

HUTCHINSON, J. (1971) Finite-Difference Solutions to Delay Differential Equations. Ph.D. thesis, Rensselair Polytechnic Inst., Troy, New York.

INCE, E.L. (1926) Ordinary Differential Equations. Dover reprint 1956.

ISAACSON, E. and KELLER, H.B. (1966) Analysis of Numerical Methods. Wiley.

JACKSON, L.W. and KENUE, S.K. (1974) A fourth order exponentially fitted method. SIAM. J. Numer. Anal. 11, 965-978.

JAIN, M.K. and SRIVASTAVA, V.K. (1970) High order stiffly stable methods for solving ordinary differential equations. Tech. Rept. No. 394. Dept. of Comput. Sci., University of Illinois.

JAIN, R.K. (1972) Some A-stable methods for stiff ordinary differential equations. Math. Comp. 26, 71-77.

JANKOWSKI, D.F., TAKEUCHI, D.I. and GERSTING, J.M. Jr. (1972) The Riccati transformation in the numerical solution of Orr-Sommerfeld problems. Trans. ASME. 94, 280.

JONES, G.S. (1962) On the non-linear differential-difference equation $f'(x) = -\alpha f(x-1)\{1+f(x)\}$. J. Math. Anal. and its Appl. 4, 440-469.

KANTOROVICH, L.V. and AKILOV, G.P. (1964) Functional Analysis in Normed Spaces, tr. D.E. Brown, Pergamon.

KELLER, H.B. (1968) Numerical Methods for Two-Point Boundary-Value Problems. Blaisdell.

KELLER, H.B. (1969) Accurate difference methods for linear ordinary differential equations subject to linear constraints. SIAM. J.

Numer. Anal. 6, 8.

KELLER, H.B. (1974) Accurate difference methods for nonlinear two-point boundary value problems. SIAM. J. Numer. Anal. 11, 305.

KELLER, H.B. (1975a) Approximation methods for nonlinear problems with application to two-point boundary value problems. Math. Comp. 29, 464-474.

KELLER, H.B. (1975b) Numerical solution of boundary-value problems for ordinary differential equations: survey and some recent results on difference methods, in AZIZ (1975). 27-88.

KELLER, H.B. and WHITE, A.B. Jr. (1975) Difference methods for boundary-value problems in ordinary differential equations. SIAM. J. Numer. Anal. 12, 791-802.

KEMPER, G.A. (1972) Linear multistep methods for a class of functional differential equations. Num. Math. 19, 361-372.

KEMPER, G.A. (1975) Spline function approximation for solutions of functional differential equations. SIAM. J. Numer. Anal. 12, 73-88.

KERSHAW, D. (1974) Volterra equations of the second kind, in DELVES and WALSH (1974). 140-161.

KLOPFENSTEIN, R.W. and MILLMAN, R.S. (1968) Numerical stability of a one-evaluation predictor-corrector algorithm for numerical solution of ordinary differential equations. Math. Comp. 22, 557-564.

KOHFELD, J.J. and THOMPSON, G.T. (1967) Multistep methods with modified predictors and correctors. J. Ass. comput. Mach. 14, 155-166.

KRAMBECK, F.J. (1970) The mathematical structure of chemical kinetics in homogeneous single-phase systems. Arch. Rat. Mech. Anal. 38, 317-347.

KREISS, H. (1972) Difference approximations for boundary and eigen-value problems for ordinary differential equations. Math. Comp. 10, 605-624.

KROGH, F.T. (1966) Predictor-corrector methods of high order with improved stability characteristics. J. Ass. comput. Mach. 13, 374-385.

KROGH, F.T. (1969) VODQ/SVDQ/DVDQ - Variable order integrators for the numerical solution of ordinary differential equations. TU Doc. No. CP-2308, NPO-11643, Jet Propulsion Lab., Pasadena, Calif.

KROGH, F.T. (1969) A variable step, variable order multistep method for the numerical solution of ordinary differential equations, in Information Processing 68 (Proceedings of the IFIP Congress 1968), North Holland Publishing Co.

KROGH, F.T. (1971) An integrator design. JPL.TM. No.33-479, Jet Propulsion Lab., Pasadena, Calif.

KROGH, F.T. (1972) Opinions on matters connected with the evaluation of programs and methods for integrating ordinary differential equations. SIGNUM newsletter 7.

KROGH, F.T. (1973a) Algorithms for changing the stepsize. SIAM. J. Numer. Anal. 10, 949-965.

KROGH, F.T. (1973b) On testing a subroutine for the numerical integration of ordinary differential equations. J. Ass. comput. Mach. 20, 545-562.

KROGH, F.T. (1974) Changing stepsize in the integration of differential equations using modified divided differences, in Lecture Notes in Mathematics. 362, Springer-Verlag. 22-71.

KUBICEK, M. and HLAVACEK, V. (1974) Evaluation of branching points based on shooting methods and GPM techniques. Chem. Eng. Sci. 29, 1695.

LAMBERT, J.D. (1973) Computational Methods in Ordinary Differential Equations. Wiley.

LAMBERT, J.D. and SIGURDSSON, S.T. (1972) Multistep methods with variable matrix coefficients. SIAM J. Numer. Anal. 9, 715-733.

LAPIDUS, L., AIKEN, R.C. and LIU, Y.A. (1974) The occurence and numerical solution of physical and chemical systems having widely varying time constants, in WILLOUGHBY (1974).

LAPIDUS, L. and SEINFELD, J.H. (1971) Numerical Solution of Ordinary Differential Equations. Academic Press.

LAWSON, J.D. and EHLE, B.L. (1972) Improved generalised Runge-Kutta, in Proc. of Canadian Computer Conference., Session 72, 223201-223213.

LENTINI, M. and PEREYRA, V. (1974) A variable order finite difference method for nonlinear multipoint boundary value problems. Math. Comp. 28, 981-1003.

LENTINI, M. and PEREYRA, V. (1975) Boundary problem solvers for first order systems based on deferred corrections, in AZIZ (1975).

LEVIN, J.J. and LEVINSON, N. (1954) Singular perturbations of non-linear systems of differential equations and an associated boundary layer equation. J. Rat. Mech. Anal. 3, 247-270.

LINDBERG, B. (1971a) On smoothing and extrapolation for the trapezoidal rule. BIT. 11, 29-52.

LINDBERG, B. (1971b) On smoothing for the trapezoidal rule, an analytical study of some representative test examples. Rept. NA.71.31., Dept. of Inf. Proc., Royal Inst. of Tech., Stockholm.

LINDBERG, B. (1972a) IMPEX - A program package for solution of systems of stiff differential equations. Rept. NA.72.50., Dept. of Inf. Proc., Royal Inst. of Tech., Stockholm.

LINDBERG, B. (1972b) Error estimates and stepsize strategy for the implicit midpoint rule with smoothing and extrapolation. Rept. NA.72.59., Dept. of Inf. Proc., Royal Inst. of Tech., Stockholm.

LINDBERG, B. (1973) IMPEX2 - A procedure for solution of systems of stiff differential equations. TRITA-NA-7303., Dept. of Inf. Proc., Royal Inst. of Tech., Stockholm.

LINDBERG, B. (1974a) On a dangerous property of methods for stiff differential equations. BIT. 14, 430-436.

LINDBERG, B. (1974b) Optimal stepsize sequences and requirements for the local error for methods for (stiff) differential equations. Tech. Rept. No.67., Dept. of Comput. Sci., University of Toronto.

LINIGER, W. (1968) A criterion for A-stability of linear multistep integration formulae. Computing. 3, 280-285.

LINIGER, W. (1969) Global accuracy and A-stability of one and two-step integration formulae for stiff ordinary differential equations, in Proc. Conf. on Numerical Solution of Differential Equations, Dundee 1968. Springer, Berlin, 188-193.

LINIGER, W. (1971) A stopping criteria for the Newton-Rpahson-method in implicit multistep integration algorithms for nonlinear systems of ordinary differential equations. Communs. Ass. comput. Mach. 14, 600-601.

LINIGER, W. (1975) Connections between accuracy and stability properties of linear multistep formulas. Communs. Ass. comput. Mach. 18, 53-56.

LINIGER, W. and WILLOUGHBY, R.A. (1970) Efficient integration methods for stiff systems of ordinary differential equations. SIAM. J. Numer. Anal. 7, 47-66.

LINZ, P. (1969) Linear multistep methods for Volterra integro-differential equations. J. Ass. comput. Mach. 16, 295-301.

MÄKELA, M., NEVANLINNA, O. and SIPILÄ, A.H. (1974) On the concepts of convergence, consistency, and stability in connexion with some numerical methods. Num. Math. 22, 261-274.

MARQUARDT, D. (1963) An algorithm for least squares estimation of nonlinear parameters. SIAM. J. Appl. Math. 11, 431.

MEYER, G.H. (1973) Initial Value Methods for Boundary Value Problems. Academic Press, New York.

MIKHLIN, S.G. (1971) The Numerical Performance of Variational Methods, trans. R. Andersson. Wolters-Noordhoff.

MIKHLIN, S.G. and SMOLITSKY, K.L. (1967) Approximate Methods for the Solution of Differential and Integral Equations. Elsevier.

MILLER, J.J.H. (1971) On the location of zeros of certain classes of polynomials with applications to numerical analysis. J. Inst. Math. and its Appl. 8, 397-406.

MITCHELL, A.R. and WAIT, R. (1976) The Finite Element Method for Partial Differential Equations. Wiley.

MIRANKER, W.L. (1973) Numerical methods of boundary alyer type for stiff systems of differential equations. Computing, 11, 221-234.

MOCARSKY, W.L. (1971) Convergence of step-by-step methods for nonlinear integro-differential equations. J. Inst. Math. and its Appl. 8, 235-239.

MURRAY, J.D. (1974) Asymptotic Analysis. Clarendon Press.

MUSA, F.A. and DELVES, L.M. (to appear) On a class of non-singular matrices. Utilitas Mathematica.

MYSKIS, A.D. (1949) General theory of differential equations with a perturbed argument. Translation: Americal Mathematical Society Translations Series 1, 4, 207-267.

NACHTSCHEIM, P.R. and SWIGERT, P. (1965) Satisfaction of Asymptotic boundary conditions in the numerical solution of systems of nonlinear equations of boundary layer type. Tech. Rept. Lewis Research Center.

NAG Mini-Manual. Mark 4. National Algorithms Group. Oxford.

NEVES, K.W. (1975) Automatic integration of functional differential equations: an approach and an algorithm. Manuscript. To appear in ACM Trans. on Mathematical Software.

NEVES, K.W. and FELDSTEIN, A. (to appear) Characterization of jump discontinuities for the state-dependent delay differential equation. J. Math. Anal. and its Appl.

NORDSIECK, A. (1962) On the numerical integration of ordinary differential equations. Math. Comp. 16, 22-49.

NORSETT, S.P. (1969a) A criterion for A(α)-stability of linear multistep methods. BIT. 9, 259-263.

NORSETT, S.P. (1969b) An A-stable modification of the Adams-Bashforth methods, in Proc. Conf. on Numerical Solution of Differential Equations, Dundee 1968. Springer, Berlin 214-219.

NORSETT, S.P. (1974) Semi explicit Runge-Kutta methods. Tech. Rept. Math. and Comp. No. 6/74, University of Trondheim.

ODÉN, L. (1971) An experimental and theoretical analysis of the SAPS-method for stiff ordinary differential equations. Rept. NA.71. 28., Dept. of Inf. Proc., Royal Inst. of Tech., Stockholm.

O'MALLEY, R.E. (1971) On initial value problems for nonlinear systems of differential equations with two small parameters. Arch. Rat. Mech. Anal. 40, 209.

OPPELSTRUP, J. (1973) Delay 1 - A program for integrating systems of delay differential equations. Rept. TRITA-NA-7311, Royal Inst. of Tech., Stockholm.

ORTEGA, J.M. and RHEINBOLDT, W.C. (1970) Iterative Solution of Nonlinear Equations in Several Variables. Academic Press.

OSBORNE, M.R. (1974) On the numerical solution of boundary value problems for ordinary differential equations, IFIPS 74. North Holland Publishing Co.

PEREYRA, V. (1966) On improving an approximate solution of a functional equation by deferred corrections. Num. Math. 8, 376-391.

PEREYRA, V. (1967) Iterated deferred corrections for nonlinear operator equations. Num. Math. 10, 316-323.

PEREYRA, V. (1968) Iterated deferred corrections for non-linear boundary value problems. Num. Math. 11, 111-125.

PEREYRA, V. (1973) Variable order variable step finite difference methods for nonlinear boundary value problems, in Lecture Notes in

Mathematics. 363. Springer-Verlag, 118-133.

PHILLIPS, G.M. (1970) Analysis of numerical iterative techniques for solving integral and integro-differential equations. Comput. J. 13, 297-300.

PINNEY, E. (1959) Ordinary Difference-Differential Equations. University of California Press, Berkeley.

POUZET, P. (1960) Méthode d'intégration numérique des equations integral et integrodifferentielles du type Volterra de seconde espese. Formules de Runge-Kutta, in Symposium on the Numerical Treatment of Ordinary Differential Equations, Integral, and Integro-Differential Equations. Rome, 1960. Birkhauser Verlag, Basel.

POUZET, P. (1962) Etude, en vue de leur traitment numerique, des équations integrales et integrodifferentielles de type Volterra pour des problèmes de conditions initiales. Thesis. University of Strasbourg.

POUZET, P. (1970) Systems differentiels, equations intégrales et integro-differentielles. (Ch. 4 of CNRS (1970).)

PROTHERO, A. and ROBINSON, A. (1974) On the stability and accuracy of one-step methods for solving stiff systems of ordinary differential equations. Math. Comp. 28, 145-162.

RALL, L.B. (1968) Davidenko's method for the solution on nonlinear algebraic equations. MRC Rept. 948. University of Wisconsin.

RHEINBOLDT, W.C. (1975) On the solution of some nonlinear equations arising in the application of finite element methods. TR 362, Comput. Sci., University of Maryland.

RILEY, J., BENNET, M. and McCORMICK, E. (1967) Numerical integration of variational equations. Math. Comp. 21, 12.

ROBERTS, S.M. and SHIPMAN, J.S. (1972a) Solution of Troesch's two-point boundary-value problem by a combination of techniques. J. Comp. Phys. 10, 232.

ROBERTS, S.M. and SHIPMAN, J.S. (1972b) Two Point Boundary Value Problems: Shooting Methods. Elsevier.

ROBERTS, S.M. and SHIPMAN, J.S. (1973) Extension of a perturbation technique for nonlinear two-point boundary-value problems. J. Opt. Theory and Appl. 12, 459.

ROBERTSON, H.H. (1966) The solution of a set of reaction rate equations, in Numerical Analysis: An Introduction, ed. J. Walsh. Academic Press.

ROBERTSON, H.H. (1967) An approximate inverse algorithm for solution of stiff differential equations. ICI Ltd. Internal Rept. MSDH/67/86.

ROBERTSON, H.H. (1975a) Numerical integration of systems of stiff ordinary differential equations with special structure. N.A. Rept. No. 8, Dept. of Math., University of Manchester.

ROBERTSON, H.H. (1975b) J. Inst. Math. and its Appl. in press.

ROBERTSON, H.H. and McCANN, M.J. (1969) A note on the numerical integration of conservative systems of first-order ordinary differential equations. Comput. J. 12, 81.

ROBERTSON, H.H. and WILLIAMS, J. (1975) Some properties of algorithms for stiff differential equations. J. Inst. Math. and its Appl. 16, 23-34.

ROSENBROCK, H.H. (1963) Some general implicit processes for the numerical solution of differential equations. Comput. J. 5, 329-330.

ROSENHEAD, L. (1963) Laminary Boundary Layers. Oxford.

ROSSER, J.B. (1967) A Runge-Kutta for all seasons. SIAM. Rev. 9, 417-452.

RUSSELL, R.D. (1975) Application of B-splines for solving differential equations. Unpublished manuscript. Simon Fraser University.

RUSSELL, R.D. and VARAH, J.M. (1974) A comparison of global methods for linear two-point boundary value problems. Tech. Rept. 74-01, Dept. of Comput. Sci., University of British Columbia.

SCHMITT, K. (ed.) (1972) Delay and Functional Differential Equations and their Applications. (Proceedings of the Park City Conference, 1972) Academic Press.

SCOTT, M. (1973a) Invariant Imbedding and its Application to Ordinary Differential Equations. Addison-Wesley.

SCOTT, M. (1973b) An initial value method for the eigenvalue problem for systems of ordinary differential equations. J. Comp. Phys. 12, 334.

SCOTT, M. (1974) A bibliography on invariant imbedding and related topics. Tech. Rept. SLA-74-0284, Sandia Laboratories.

SCOTT, M., SHAMPINE, L.F. and WING, G.M. (1969) Invariant imbedding and the calculation of eigenvalues for Sturm-Liouville systems. Computing, 4, 10.

SCOTT, M. and WATTS, H.A. (1975) SUPORT - A computer code for two-point boundary value problems via orthonormalisation. Rept. SAND 75-0198, Sandia Laboratories.

SCRATON, R.E. (1964) Estimation of the truncation error in Runge-Kutta and allied processes. Comput. J. 7, 246-248.

SEDGWICK, A. (1973) An effective variable order variable step Adams method. Tech. Rept. No. 53., Dept. of Comput. Sci., University of Toronto.

SHAMPINE, L.F. (1971) Comparing error estimates for Runge-Kutta methods. Math. Comp. 25, 445-455.

SHAMPINE, L.F. (1973) Local extrapolation in the solution of ordinary differential equations. Math. Comp. 27, 91-97.

SHAMPINE, L.F. and GORDON, M.K. (1975) Computer Solution of Ordinary Differential Equations. W.H. Freeman and Co.

SHAMPINE, L.F. and WATTS, H.A. (1969) Block implicit one-step methods. Math. Comp. 23, 731-740.

SHAMPINE, L.F. and WATTS, H.A. (1972) A-stable block implicit one-step methods. BIT. 12, 252-266.

SHINTANI, H. (1966a) On a one-step method of order 4. J. Sci. Hiroshima Univ. Ser. A-1 Math. 30, 91-107.

SHINTANI, H. (1966b) Two-step processes by one-step methods of order 3 and of order 4. Ibid., 183-195.

SHORT, E. (1975) On the Stability of some Numerical Methods for the Solution of Initial-Value Problems in Volterra Integro-Differential Equations. M.Sc. thesis, University of Manchester.

SOOP, M. (1968) A method for calculating the minimum of a one-dimensional variational integral on an infinite interval leading to a Schrödinger type eigenvalue problem. ZAMM. 48, 369.

SPIJKER, M.N. (1966) Convergence and stability of step-by-step methods for the numerical solution of initial value problems. Num. Math. 8, 161-177.

STETTER, H.J. (1965) A study of strong and weak stability in discretization algorithsm. SIAM. J. Numer. Anal. 2, 265-280.

STETTER, H.J. (1971) Stability of discretizations on infinite intervals, in Lecture Notes in Mathematics. 228, Springer-Verlag. 207-222.

STETTER, H.J. (1973) Analysis of Discretization Methods for Ordinary Differential Equations. Springer-Verlag.

STOCKMAYER, W.H. (1944) The steady-state approximation in polymerisation kinetics. J. Chem. Phys. 12, 143-144.

STOER, J. (1961) Uber zwei Algorithmen zur Interpolation mit rationalen funktionen. Num. Math. 3, 285-304.

STOER, J. (1974) Extrapolation methods for the solution of initial value problems and their practical realisation, in Lecture Notes in Mathematics. 362, Springer-Verlag. 1-21.

STRANG, G. and FIX, G. (1973) An Analysis of the Finite Element Method. Prentice-Hall.

SZEWCZYK, A.A. (1964) Combined forced and free-convection laminar flow. J. Heat Transfer. Trans. A.S.M.E. 86, 501.

TAUBERT, K. (1974) Differenzenverfahren fur gewonlicher Anfangswertaufgaben mit unstetiger rechter Seite. Rept. LN-395, Institute fur Angewandte Mathematik, Hamburg.

TAVERNINI, L. (1969) Numerical Methods for Volterra Functional Differential Equations. Ph.D. thesis, University of Wisconsin at Madison.

TAVERNINI, L. (1971) One-step methods for the numerical solution of Volterra functional differential equations. SIAM. J. Numer. Anal. 8, 786-795.

TAVERNINI, L. (to appear) Linear multistep methods for the numerical solution of Volterra functional differential equations. J. of Applic. Anal.

THOMEE, V. and WAHLBIN, L. (1975) On Galerkin methods for semilinear parabolic problems. SIAM. J. Numer. Anal. 12, 378-389.

THOMPSON, R.J. (1968) On some functional differential equations: existence of solutions and difference approximations. SIAM. J. Numer. Anal. 5, 475-487.

TREANOR, C.E. (1966) A method for the numerical integration of coupled first-order differential equations with greatly different time constants. Math. Comp. 20, 39-45.

TSOI, A.C. (1975) Explicit solution of a class of delay-differential equations. International J. of Control. 21, 39-48.

VAN de LUNE, J. and WATTEL, E. (1969) On the numerical solution of a differential-difference equation arising in analytic number theory. Math. Comp. 23, 417-421.

VARAH, J.M. (1972) On the solution of block-tridiagonal systems arising from certain finite-difference equations. Math. Comp. 26, 859.

VARAH, J.M. (1973) A comparison of some numerical methods for two-point boundary value problems. Tech. Rept. 73.01. Dept. of Comput. Sci., University of British Columbia.

VASIL'EVA, A.B. (1963) Asymptotic behaviour of solutions of certain problems for ordinary differential equations with a small parameter multiplying the highest derivatives. Usp. Nat. Nauk. (Russian) 18

WAIT, R. and HOPKINS, T.R. (1975) A comparison of numerical methods for solving parabolic partial differential equations, in Proc. of Conf. on Computational Aspects of Finite Element Methods, Imperial College.

WALSH, J. (1974) Initial and boundary value routines for ordinary differential equations, in Proc. of the Conf. on Software for Numerical Mathematics 1973, Academic Press.

WASSERSTROM, E. (1973) Numerical solutions by the continuation method. SIAM. Rev. 15, 89.

WASOW, W. (1965) Asymptotic Expansions for Ordinary Differential Equations. Interscience, New York.

WATT, J.M. (1967) The asymptotic discretization error of a class of methods for solving differential equations. Proc. Camb. Phil. Soc. 63, 461-472.

WATT, J.M. (1968) Convergence and stability of discretization methods for functional equations. Comput. J. 11, 77-82.

WEI, J. (1962) Axiomatic treatment of chemical reaction systems. J. Chem. Phys. 36, 1578.

WEI, J. and PRATER, C.D. (1962) Analysis of complex reaction systems. Adv. in Catalysis. 13, 203-391.

WEISS, R. (1974) The application of implicit Runge-Kutta and collocation methods to boundary-value problems. Math. Comp. 28, 449.

WHEELER, D.J. (1959) Note on Runge-Kutta method of integrating ordinary differential equations. Comput. J. 2, 23.

WHYTACK, L. (1971) A new technique for rational extrapolation to the limit. Num. Math. 17, 215-221.

WIDLUND, O.B. (1967) A note on unconditionally stable linear multistep methods. BIT. 7, 65-70.

WIEDERHOLT, L.F. (1970) Numerical Integration of Delay-Differential Equations. Ph.D. thesis, University of Wisconsin.

WILKINSON, J.H. (1965) The Algebraic Eigenvalue Problem. Clarendon Press, Oxford.

WILLIAMS, J. and de HOOG, F. (1974) A class of A-stable advanced multistep methods. Math. Comp. 28, 163-177.

WILLOUGHBY, R.A. (ed.) (1974) Stiff Differential Systems. Plenum Press.

ZAFARULLAH, A. (1970) Application of the method of lines to parabolic partial differential equations with error estimates. J. Ass. comput. Mach. 17, 294-302.

ZDANOV, G.M. (1961) On approximate solution of a system of differential equations of first order with retardation. Uspekhi Mat. Nauk. 16.

ZVERKIN, A.M., KEMENSKI, G.A., NORKIN, S.B. and EL'SGOL'TS, L.E. (1962) Differential equations with a perturbed argument. Russian Math. Surveys. 17, 61-146.

ZVERKINA, T.S. (1964) A modification of finite difference methods for integrating ordinary differential equations with non-smooth solution. Vyzisl. Mat. i Mat. 4(4) Supplement, 149-160. MR35 No.2492.

ZVERKINA, T.S. (1966) A new class of finite-difference operators. Sov. Math. 7(6), 1412-1415.

ZVERKINA, T.S. (1968) A one-parameter analogue of Adams' formulae. USSR Comp. Math. and Math. Phys. 8, 124-139.

ZVERKINA, T.S. (1962) Approximate solution of differential equations with perturbed argument and of differential equations with discontinuous right-hand sides. (In Russian.) Trudy Sem. Teor. Differencial. Uravensii s Otklon. Argumentom, Univ. Druzby Naradov Patrisa Lumumby 1, 76-93. MR32. No.2708.

ZVERKINA, T.S. (1965) Modified Adams formula for integrating equations with deviating argument. (In Russian.) Ibid. 3, 221-232. MR35. No.5131.

ZVERKINA, T.S. (1967) Numerical integration of systems with time lag. (In.Russian.) Ibid. 4, 164-172. MR36. No.7332.

ZVERKINA, T.S. (1967) A new class of finite operators. (In Russian.) Ibid. 5, 90-106, MR37. No.628.

SUBJECT INDEX